REGULATION OF CARBON PARTITIONING IN PHOTOSYNTHETIC TISSUE

edited by

ROBERT L. HEATH

and

JACK PREISS

PROCEEDINGS OF THE EIGHTH ANNUAL
SYMPOSIUM IN PLANT PHYSIOLOGY
(January 11-12, 1985)
UNIVERSITY OF CALIFORNIA, RIVERSIDE

Proceedings sponsored by
the Department of Botany and Plant Sciences, UCR;
Dean, College of Natural and Agricultural Sciences, UCR;
Chancellor's Office, UCR;
Director, California Agricultural Experiment Station, UC;
National Science Foundation; and
the U.S. Department of Energy.
Assistance was also provided by Beckman Instruments, Inc.
and Calgene, Inc.

Copyright © 1985
American Society of Plant Physiologists
15501-A Monona Drive
Rockville, Maryland 20855

All rights reserved. No part of this publication may be reproduced without the prior written permission of the publisher.

Symposium in Plant Physiology (8th : 1985 : University of California, Riverside)
    Regulation of Carbon Partitioning in Photosynthetic Tissue

"Proceedings sponsored by the University of California, Riverside, with support from the National Science Foundation, the U. S. Department of Energy, and the Director of the California Agricultural Experiment Station."

Library of Congress Cataloging in Publication Data
Symposium in Plant Physiology (8th : 1985 : University
    of California, Riverside)
    Regulation of carbon partitioning in photosynthetic
tissue.

    "Proceedings sponsored by the Department of Botany
and Plant Sciences, UCR"... [et al.]
    Includes index.
    1. Photosynthesis--Congresses. 2. Carbon--Metabolism
--Congresses. 3. Plant translocation--Congresses.
I. Heath, Robert L., 1940-   . II. Preiss, Jack,
1932-   . III. University of California, Riverside.
Dept. of Botany and Plant Sciences. IV. Title.

QK882.S944 1985   581.1'3342   85-13473
ISBN 0-943088-07-0

Printed at the Waverly Press, Baltimore, Maryland 21202

Front Row: Preiss, Beevers, Black, Gibbs, Heath, Laties, Walker, Loescher (left to right)

Middle Row: ap Rees, Thomson, Outlaw, Giaquinta, Knowles, Portis, Wyse, Lucas, Geiger, Fondy, Ting

Back Row: Stitt, Latzko, Seftor, Gustafson, Dennis, Beck, Huber, Robinson, Hendrix, Koch, Humphreys

CONTRIBUTORS

BENY ALONI. Plant Biochemistry Laboratory, Utah State University, Logan, UT 84322 USA.

THOMAS AP REES. Botany School, University of Cambridge, Downing Street, Cambridge CB2 3EA, England.

CHRIS BAYSDORFER. Plant Photobiology Laboratory, U. S. Department of Agriculture, BARC-W, Beltsville, MD 20705 USA.

ERWIN BECK. Lehrstuhl Pflanzenphysiologie, University of Bayreuth, Universitatsstrasse 30, D-8580 Bayreuth, West Germany.

HARRY BEEVERS. Biology Department, University of California, Santa Cruz, CA 95064 USA.

WILLIAM E. BELKNAP. Department of Molecular Genetics and Cell Biology, University of Chicago, Chicago, IL 60637 USA.

CLANTON C. BLACK. Department of Biochemistry, University of Georgia, Athens, GA 30602 USA.

FREDERIK C. BOTHA. Department of Biology, Queen's University, Kingston ON K7L 3N6, Canada.

DONALD BRISKIN. Plant Biochemistry Laboratory, Utah State University, Logan, UT 84322 USA.

NANCY W. CARNAL. Department of Biochemistry, University of Georgia, Athens, GA 30602 USA.

JEANINE M. DAVIS. Horticultural Department, Washington State University, Pullman, WA 99164 USA.

DAVID T. DENNIS. Department of Biology, Queen's University, Kingston ON K7L 3N6, Canada.

Speakers and poster participants at Symposium

JEFFREY EDWARDS. Botany School, University of Cambridge, Downing Street,
   Cambridge CB2 3EA, England.
JOHN K. FELLMAN. Horticultural Department, Washington State University,
   Pullman, WA 99164 USA.
*BERNADETTE R. FONDY. Department of Biology, Seton Hill College, Greensburg, PA
   15601 USA.
TED C. FOX. Horticultural Department, Washington State University,
   Pullman, WA 99164 USA.
CHEE FOOK FU. Institute for Photobiology of Cells and Organelles, Brandeis
   University, Waltham, MA 02254 USA.
*DONALD R. GEIGER. Department of Biology, University of Dayton, College Park,
   Dayton, OH 45469 USA.
KLAUS-PETER GERBLING. Botanisches Institut, Schlossgarten 3, Munster, 74400,
   West Germany.
*MARTIN GIBBS. Institute for Photobiology of Cells and Organelles, Brandeis
   University, Waltham, MA 02254 USA.
JOHN H. GREEN. Botany School, University of Cambridge, Downing Street,
   Cambridge CB2 3EA, England.
*STEVEN W. GUSTAFSON. Department of Biochemistry, University of Arizona, Tucson
   AZ 85721 USA.
*ROBERT L. HEATH. Department of Botany and Plant Sciences, University of
   California, Riverside, CA 92521 USA.
WILMA E. HEKMAN. Department of Biology, Queen's University, Kingston ON K7L 3N
   Canada.
*JOHN E. HENDRIX. Department of Botany, Colorado State University, Fort Collins
   CO 80523 USA.
*STEVEN C. HUBER. U. S. Department of Agriculture, Agricultural Research Service
   Crop Science Department, North Carolina State University, Raleigh, NC 27606
   USA.
*THOMAS E. HUMPHREYS. Vegetable Crops Department, University of Florida,
   Gainesville, FL 32611 USA.
ROBERT J. IRELAND. Department of Biology, Queen's University, Kingston
   ON K7L 3N6, Canada.
LEANNE M. JABLONSKI. Department of Biology, University of Dayton, College Park
   Dayton, OH 45469 USA.

RICHARD G. JENSEN. Department of Biochemistry, University of Arizona, Tucson, AZ 85721 USA.

WILLY KALT-TORRES. U. S. Department of Agriculture, Agricultural Research Service, Crop Science Department, North Carolina State University, Raleigh, NC 27606 USA.

PHILLIP S. KERR. U. S. Department of Agriculture, Agricultural Research Service, Crop Science Department, North Carolina State University, Raleigh, NC 27606 USA.

ROBERT A. KENNEDY. Horticultural Department, Washington State University, Pullman, WA 99164 USA.

*FRANCIS C. KNOWLES. Marine Biology Research Division, A-002, Scripps Institute of Oceanography, University of California-San Diego, La Jolla, CA 92093 USA.

*KAREN E. KOCH. Department of Fruit Crops, University of Florida, Gainesville, FL 32611 USA.

NICHOLAS J. KRUGER. Department of Biology, Queen's University, Kingston ON K7L 3N6, Canada.

*ERWIN LATZKO. Botanisches Institut, Schlossgarten 3, Munster, 74400, West Germany.

*WAYNE LOESCHER. Horticultural Department, Washington State University, Pullman, WA 99164 USA.

*WILLIAM J. LUCAS. Department of Botany, University of California, Davis, CA 95616 USA.

KATHLEEN McNAMARA. Department of Biochemistry and Biophysics, University of California, Davis, CA 95616

SUSAN MORRELL. Department of Biochemistry, University of Georgia, Athens, GA 30602 USA.

*WILLIAM H. OUTLAW, JR. Biological Science Department, Florida State University, Tallahassee, FL 32306 USA.

NACHMAN PAZ. Department of Biochemistry, University of Georgia, Athens, GA 30602 USA.

BERNARD J. PLOEGER. Department of Biology, University of Dayton, College Park, Dayton, OH 45469 USA.

*ARCHIE R. PORTIS, JR. U. S. Department of Agriculture, Department of Agronomy, University of Illinois, Urbana, IL 61801 USA.

*JACK PREISS. Department of Biochemistry, Michigan State University, East Lansing, MI 48824 USA. (formerly of Department of Biochemistry and Biophysics, University of California, Davis, CA 95616 USA)

PATRICIA G. RAY. AG TSD, Dow Chemical USA, 2800 Mitchell Drive, Walnut Creek, CA 94598 USA.

DEBORAH A. RAYNES. Department of Biochemistry, University of Arizona, Tucson, AZ 85721 USA.

ROBERT J. REDGWELL. Division of Horticulture and Processing, DSIR, Private Bag, Auckland, New Zealand.

*J. MICHAEL ROBINSON. Plant Photobiology Laboratory, U. S. Department of Agriculture, BARC-W, Beltsville, MD 20705 USA.

NINA ROBINSON. Department of Biochemistry and Biophysics, University of California, Davis, CA 95616 USA.

*RICHARD E. B. SEFTOR. Department of Biochemistry, University of Arizona, Tucson, AZ 85721 USA.

MIRTA N. SIVAK. Research Institute for Photosynthesis, Department of Botany, University of Sheffield, Sheffield S10 2TN, United Kingdom.

STEVEN SPILATRO. Department of Biochemistry, Michigan State University, East Lansing, MI 48824 USA.

STEFAN A. SPRINGER. Biological Science Department, Florida State University, Tallahassee, FL 32306 USA.

MARTIN STEUP. Botanisches Institut, Schlossgarten 3, Munster, 74400, West Germany.

*MARK STITT. Institute Fur Biochemie der Pflanze, University of Gottingen, Untere Karspule, Gottingen D-3400, West Germany.

MITCHELL C. TARCZYNSKI. Biological Science Department, Florida State University Tallahassee, FL 32306 USA.

ALAN THOMSON. Department of Biology, Queen's University, Kingston ON K7L 3N6, Canada.

*DAVID A. WALKER. Research Institute for Photosynthesis, Department of Botany, University of Sheffield, Sheffield S10 2TN, United Kingdom.

PATRICIA M. WILSON. Botany School, University of Cambridge, Downing Street, Cambridge CB2 3EA, England.

*ROGER E. WYSE. Plant Biochemistry Laboratory, Utah State University, Logan, UT 84322 USA.

EDITORS' INTRODUCTION

Most researchers in Plant Physiology have come to realize that in order to increase the yield of agriculturally important crops, the processes of photosynthesis and translocation have to be understood. Once the basic mechanisms are firmly elucidated, geneticists can then try to optimize the interactions between the two processes controlling the movement of the newly fixed carbon throughout the plant.

In the seventies, great strides were made in comprehending the regulation occurring within the photosynthetic enzymes, critical in the control of the flow of carbon from the first photosynthetic product to the more complex carbohydrates which serve as the carbon-transporting agents. While the mechanisms of translocation remain partially obscured, various model systems have been developed over the past decade which should allow researchers a better view over the barriers.

The real stimulus for organizing this Symposium--held as the Eighth Annual Symposium in Plant Physiology and Botany at the University of California, Riverside, on January 11 and 12, 1985--was the discovery of a critical carbohydrate of metabolism, fructose-2,6-bisphosphate. Shortly after its discovery, many laboratories realized that this metabolite was a key that would unlock many doors leading to the understanding of regulation and control of the flow of carbon in photosynthetically active leaves. Although all the rooms have yet to be explored, it is clear that a great new area has been opened which will lead to a much clearer understanding of the movement of carbon throughout the plant. The scientific community is now into the exploratory phase and we, the Editors, hope that this volume will help all students and researchers. To that end, we have tried to release this book rapidly and keep the cost low so that the students can easily possess, read, and re-read the ideas written by the leaders in this field.

The participants in the Symposium were asked to give papers according to the areas in which they were currently publishing. When arranging the manuscripts of the participants, after the speeches of the Symposium were given, it seemed logical to us to group them in the order presented here. No doubt, others would group the papers differently, yet we hope the logic behind this order will surface.

The fundamental principles of photosynthesis are assumed: light is trapped within the grana to produce NADPH and ATP which, in turn, are used to fix carbon dioxide and produce 3-phosphoglyceric acid (PGA). PGA is already formed at the beginning of our discussions. Thus, this volume begins with starch metabolism (Preiss et al. and Beck), since starch is the cellular storage material from which carbon is taken to other plant tissues and to which carbon is added by photosynthesis.

The volume then moves into the examination of the enzymes which transform PGA into the many other compounds of carbohydrate metabolism. These varied enzymes, with their regulation mechanisms, are discussed by Black et al., Latzko et al., and ap Rees et al. Here, we find the complex story of activators, inhibitors, and modulators which can change most, if not all, of the enzymatic properties. These are the enzymes thought to be the control points of carbohydrate metabolism.

Much of the control of carbohydrate metabolism is carried out by the membranes separating the various compartments within the cell. This area is well discussed by the next series of papers. Walker and Sivak investigate the use of fluorescence to study nondestructively the flow of metabolites between compartments. The papers by Stitt and by Dennis et al. discuss the role of compartments in controlling metabolite levels by fructose-2,6-bisphosphate and by isozymes located in diverse compartments. The final paper by Portis et al. describes how the phosphate transporter located on the chloroplast envelope may possibly control levels of the metabolites within and outside the chloroplast. This is the point where a discussion of cell isolation and the characterizations of the enzymes within a single cell by Outlaw et al. seemed to be most appropriate.

Our discussions of <u>Cellular Partitioning</u> end with several short papers, which were presented as posters within the Symposium. Carbon fixation through ribulose-1,6-bisphosphate carboxylase/oxygenase is discussed by Gustafson <u>et al</u>. and Seftor and Jensen. Knowles adds to our understanding of mechanisms with a discussion of intermediates of several enzymes within the Calvin cycle. Wu and Gibbs complete the picture by a discussion of how temperature can affect photosynthesis.

The volume then moves into <u>Tissue Partitioning</u> and how the carbon can flow into the major transport carbohydrate, sucrose. Huber <u>et al</u>. present a description of how cells make sucrose and what events can alter that production. The transport of sucrose is discussed by Humphreys in terms of uptake by the sink tissues, by Wyse <u>et al</u>. in terms of control by organelles within the uptake cells, by Lucas who suggests that for all our knowledge, we need new systems and much better understanding of phloem transport. Yet, during the translocation of sucrose, many dramatic transformations can occur which can alter the simple picture, as illustrated by Hendrix's paper.

Geiger <u>et al</u>. attempt to place the source and sink behavior into a model for the whole plant. Loescher <u>et al</u>. remind us, lest we focus too heavily on sucrose, that many other types of carbohydrates are transported. Finally, Robinson and Baysdorfer open our thinking to the interactions of carbon metabolism with nitrogen metabolism, which is critical for proper plant growth.

This section is completed by a collection of shorter papers. Fondy and Geiger discuss the control of carbohydrate flow during the day. Koch discusses sink relations in a citrus system which has potential for clarifying sink behavior.

Professor Beevers had the difficult job of summarizing the entire Symposium and giving an overview of the fields. His pointed remarks remind us that we have come far, yet have far to go. And so the end of our Symposium was reached. We feel that the papers were excellent and indicative of the progress made in this field.

No symposium can be run and a volume published without the aid of many people and institutions. This one is no exception. We would like to thank the financial support given to the Symposium by the National Science Foundation, the Department of Energy, the California Agricultural Research Station, Beckman Instruments, and Calgene. UC Riverside has provided excellent support through

Vice-Chancellor T. Hullar, Dean I. Sherman, and the Department of Botany and Plant Sciences. Finally, the Symposium could not have been run without the excellent and competent organization of Cindi McKernan. We also greatly appreciate the time of Drs. I. P. Ting and W. W. Thomson, both of UCR, and Dr. G. Laties of UCLA, who acted as moderators during the Symposium. The Symposium could not have been published without the fine typing and proofreading of Patti Garcia and Aileen Wietstruk, the proofreading and the indexing of Paula Frederick, and the photographs by Mike Elderman. Of course, as always, the Editors must take the final blame for errors in the duplication of the manuscripts; our pens were the last on the paper.

> Robert L. Heath
> Jack Preiss

FOREWORD

This volume represents a continuation of the series of publications on timely topics initiated recently by the American Society of Plant Physiologists. Publication of symposia devoted to focus areas, such as the present one on recent research on photosynthesis and translocation, is designed to share the information from such proceedings with other plant scientists. It is the wish of the Publications Committee and the Executive Committee to make these publications as useful as possible and, to this end, suggestions will always be welcome.

> The ASPP Publication Committee
>
> Elisabeth Gantt, Chair
> Robert Rabson
> Jack C. Shannon
> Lowell D. Owens
> Machi F. Dilworth

CONTENTS

Contributors v
Editors' Introduction ix
    R. L. HEATH and JACK PREISS
Foreword. xiii
Contents. xiv
List of Abbreviations xvii

                              *   *   *

Cellular Partitioning
- Starch Synthesis and Its Regulation 1
    JACK PREISS, NINA ROBINSON, STEVEN SPILATRO, and KATHLEEN McNAMARA
- The Degradation of Transitory Starch Granules in Chloroplasts 27
    ERWIN BECK
- Roles of Pyrophosphate and Fructose-2,6-Bisphosphate in Regulating
  Plant Sugar Metabolism 45
    CLANTON C. BLACK, NANCY W. CARNAL, AND NACHMAN PAZ
- Characterization of Fructose-1,6-Bisphosphatase Activity from
  Synechococcus leopoliensis 63
    ERWIN LATZKO, KLAUS-PETER GERBLING, and MARTIN STEUP
- Pyrophosphate and the Glycolysis of Sucrose in Higher Plants 76
    TOM AP REES, SUSAN MORRELL, JEFFREY EDWARDS, PATRICIA M. WILSON,
    and JOHN H. GREEN
- In vivo Chlorophyll a Fluorescence Transients Associated with Changes
  in the CO_2 Content of the Gas-Phase 93
    DAVID A. WALKER and MIRTA N. SIVAK

- Fine Control of Sucrose Synthesis by Fructose-2,6-Bisphosphate 109
    MARK STITT
- Compartmentation of Glycolytic Enzymes in Plant Cells 127
    DAVID T. DENNIS, WILMA E. HEKMAN, ALAN THOMSON, ROBERT J. IRELAND,
    FREDERIK C. BOTHA, and NICHOLAS J. KRUGER
- Metabolite Levels, Chloroplast Envelope Transport, and Chloroplast
    Metabolism 147
    ARCHIE R. PORTIS, JR., PATRICIA G. RAY, and WILLIAM E. BELKNAP
- Enzyme Assays at the Single-Cell Level: Real-time, Quantitative, and
    Using Natural Substrate in Solution 162
    WILLIAM H. OUTLAW, JR., STEFAN A. SPRINGER, and
    MITCHELL C. TARCZYNSKI
- Carboxylase Response to CO_2 and O_2 in Intact Leaves of Wheat and
    Maize 180
    STEVEN W. GUSTAFSON, DEBORAH A. RAYNES, and RICHARD G. JENSEN
- Chloroplasts at Air Levels of CO_2: Factors Influencing
    CO_2-fixation 184
    RICHARD E. B. SEFTOR and RICHARD G. JENSEN
- Regulation of Resynthesis of the Photosynthetic CO_2-acceptors:
    Feedback Inhibition of Transketolase 188
    FRANCIS C. KNOWLES
- CO_2 Photoassimilation of Heat-Stressed Spinach Chloroplasts 192
    CHEE F. FU and MARTIN GIBBS

Tissue Partitioning
- Regulation of Sucrose Formation and Movement 199
    STEVEN C. HUBER, PHILLIP S. KERR, and WILLY KALT-TORRES
- The Influence of External pH on Sucrose Uptake and Release in the
    Maize Scutellum 215
    THOMAS E. HUMPHREYS
- Sucrose Transport: Regulation and Mechanism at the Tonoplast 231
    ROGER WYSE, DONALD BRISKIN, and BENY ALONI
- Phloem-Loading: A Metaphysical Phenomenon? 254
    WILLIAM J. LUCAS
- Dedication to Swanson 272
- Partitioning of Sucrose into Fructans in Winter Wheat Stems 273
    JOHN E. HENDRIX

- Significance of Carbon Allocation to Starch in Growth of *Beta vulgaris* L. 289
  DONALD R. GEIGER, LEANNE M. JABLONSKI, and BERNARD J. PLOEGER
- Other Carbohydrates as Translocated Carbon Sources: Acyclic Polyols and Photosynthetic Carbon Metabolism 309
  WAYNE H. LOESCHER, JOHN K. FELLMAN, TED C. FOX, JEANINE M. DAVIS, ROBERT J. REDGWELL, and ROBERT A. KENNEDY
- Interrelationships Between Photosynthetic Carbon and Nitrogen Metabolism in Mature Soybean Leaves and Isolated Leaf Mesophyll Cells 333
  J. MICHAEL ROBINSON and CHRIS BAYSDORFER
- Diurnal Changes in Carbon Allocation: Morning 358
  BERNADETTE R. FONDY and DONALD R. GEIGER
- Nonvascular Transfer of Assimilates in Citrus Juice Tissues 362
  KAREN E. KOCH
- Regulation of Carbohydrate Partitioning 367
  HARRY BEEVERS

Index 370

LIST OF ABBREVIATIONS

Abbreviation	Name
$A_{0.5}$	concentration of activator of an enzyme giving 50% of maximal activation
ADPG	ADP-glucose
BS	bis(sulfosuccinimidyl)suberate
CCCP	mCl-carbonyl cyanide phenyl-hydrazone
CE	carboxylation efficiency
"CO_2"	CO_2 + HCO_3^-, total carbon that is in solution
D enzyme	debranching enzyme for starch glucan (donor) + glucan (acceptor)--->glucan (donor + acc) + glucose (free)
DAP	dihydroxyacetone-3-phosphate, usually DHAP
Δp	proton motive force (chemical potential of proton gradient/Faraday's constant)

DHAP	dihydroxyacetone-3-phosphate (sometimes listed as DAP)
DNP	2,4 dinitrophenol
ELISA	enzyme-linked immunoadsorbent assay
Epps	N-(2-hydroxyethyl)-piperazine-N'-3-propane sulfonic acid
FbPase	fructose-1,6-bisphosphate,1-phosphatase (E.C. 3.1.3.11)
FFTase	β-(2-1)-fructan:β-(2-1)-fructan-1-fructosyl transferase (glc-fru-fru + glc-fru-fru---sucrose + glc-(fru)$_3$)
Fru-1,6-P$_2$	fructose-1,6-bisphosphate
Fru-2,6-P$_2$	fructose-2,6-bisphosphate
Fru6-P	fructose-6-phosphate
Fru6P,2kinase	fructose-6-phosphate, 2-kinase (ATP+Fru6P---Fru-2,6-P$_2$ + ADP)
HPI	hexose-6-phosphate isomerase
$I_{0.5}$	concentration of inhibitor of an enzyme yielding 50% inhibition
kat	unit of enzyme activity, mol of product/min, (kat=10^6 U)
Ki	concentration of inhibitor required for 50% inhibition of an enzyme reaction
LD	long day (plants grown under a long-day protocol)

Mops	3-(N-morpholino)-propane-sulfonic acid
MPRase	mannose-6-phosphate reductase
N	combined nitrogen levels in nutrient solution (NH_4 + NO_3)
n	Hill constant; slope of line on Hill plot
NCE	net carbon exchange
NSC	nonstructural carbohydrate
pCMBS	p-chloro-mercuribenzene sulfonate
PEP	phosphoenolpyruvate
PFKase	ATP-dependent phosphofructokinase (sometimes listed as PFK)
PFPase	PPi-dependent phosphofructokinase (phosphofructophosphotransferase) (sometimes listed as PFP)
PGA	3-phosphoglyceric acid (when unnumbered, the 3-isomer is meant)
PMF	proton motive force (chemical potential of proton gradient and membrane potential)
PPFD	photosynthetically-active photon flux density

PPi	pyrophosphate
ψ	electrical membrane potential
PVP	polyvinyl polypyrrolidone
Q, qQ, qe	fluorescence quencher molecule at reducing end of Photosystem II (qQ = Duysens-Sweers quencher, reduction of Q; qe = Krause quencher, H^+ gradient dependent)
R enzyme	amylopectin or β-limit dextran isoamylase (possible debranching enzyme)
Rm	mesophyll resistance
Rubisco	ribulose-1,5-bisphosphate carboxylase/oxygenase (E.C. 4.1.1.39)
RuBP	ribulose-1,5-phosphate
$S_{0.5}$	concentration of substrate required for an enzyme velocity of 50% of maximum velocity, may be the same as Km
SD	short day (plants grown under a short-day protocol)
se-cc	sieve element-companion cell
SLA	specific leaf area
SPSase	sucrose phosphate synthetase (E.C. 2.4.1.14)

SSase	sucrose synthase (E.C. 2.4.1.13)
SSTase	sucrose-sucrose fructosyl transferase (2-sucrose--->glc-fru-fru + glc)
TPMP	triphenyl methyl phosphonium ion
U	units of enzyme activity, μmol of product/min (see kat), $U = 10^{-6}$ Kat
UDPG	UDP-glucose

All remaining abbreviations are according to those found in <u>Plant Physiology</u>, January 1985.

Other Publications of the American Society of Plant Physiologists

CRASSULACEAN ACID METABOLISM
Ting and Gibbs

BIOSYNTHESIS AND FUNCTION OF PLANT LIPIDS
Thompson, Mudd and Gibbs

STRUCTURE, FUNCTION, AND BIOSYNTHESIS OF PLANT CELL WALLS
Dugger and Bartnicki-Garcia

EXPLOITATION OF PHYSIOLOGICAL AND GENETIC VARIABILITY TO ENHANCE CROP PRODUCTIVITY
Harper, Schrader and Howell

Starch Synthesis and Its Regulation

JACK PREISS, NINA ROBINSON, STEVEN SPILATRO, and KATHLEEN McNAMARA

In many leaves, starch is a common product of photosynthetic CO_2 fixation and in many storage tissues is the major, if not sole, carbon and energy reserve product.

It is found in all organs of most higher plants, and invariably in higher plant cells it is synthesized in plastids. The starch granules in leaves are synthesized in the chloroplast, while those found in reserve tissue (e.g. maize endosperm; potato tuber) are generally located in the amyloplast.

The mechanism for synthesis of starch either in the leaf or non-photosynthetic tissue is quite similar and is believed to occur by three reactions. First, synthesis of the glucosyl donor, ADP-glucose, is catalyzed by ADP-glucose pyrophosphorylase (E.C.2.7.7.27; reaction 1). The glucosyl unit of ADP-glucose and, possibly to a small extent, of UDP-glucose, is then transferred to the nonreducing end of an α-glucan primer to form a new α-1,4-glucosidic linkage. This reaction is catalyzed by starch synthase (E.C.2.4.1.2.1; reaction 2). Finally, formation of the α-1,6 linkages found in amylopectin is catalyzed by branching enzyme (E.C.2.4.1.2.4; reaction 3).

$$\text{ATP} + \alpha\text{-glucose-1-P} \rightleftharpoons \text{ADP-glucose} + \text{PPi} \tag{1}$$
$$\text{ADP-glucose} + \alpha\text{-glucan} \longrightarrow \alpha\text{-1,4-glucosyl-}\alpha\text{-glucan} + \text{ADP} \tag{2}$$
$$\text{Linear } \alpha\text{-glucan chain} \longrightarrow \text{branched } \alpha\text{-1,6 - }\alpha\text{-1,4-glucan} \tag{3}$$

The properties of these enzymes from a number of sources have been reported and extensively reviewed (42, 43, 46, 48). However, the exact mechanism of formation of the starch granule is not completely understood (see Beck, this symposium). The process of starch granule formation appears to involve multiple forms of starch synthase (Rx.2) and branching enzyme (Rx.3) (5, 6, 19, 20, 30, 38-41). Moreover, although the regulatory mechanisms of starch synthesis appear to be similar for different cell types, it is not completely understood how these mechanisms are coordinated with the specialized physiology and chemistry of the different plant cell types. This report will thus present recent information on these aspects of starch synthesis.

CHARACTERIZATION OF THE MULTIPLE FORMS OF STARCH SYNTHETASE AND BRANCHING ENZYME

In many plant tissues, photosynthetic or nonphotosynthetic starch synthase activity found in the plastids is either in the soluble phase or is firmly associated with the starch granules. Partial purification of the soluble starch synthase activity indicated at least two enzyme forms. Recently, the starch-granule-bound starch synthase activity of maize endosperm has been solubilized and indeed the activity could be resolved into two protein peaks by anion exchange chromatography (30). The two peaks of activity seen in maize endosperm, either in the soluble or granule-bound phase, can be grouped into two classes that are distinguished by their kinetic and physical properties, Type I and Type II (47). Type I enzyme elutes from DEAE-sepharose and aminobutyl sepharose at lower salt concentration. Moreover, Type I starch synthases have lower mol wt (72 kDa) and show higher activities with glycogen than with amylopectin in contrast to the Type II starch synthases (mol wt = 95 kDa; 20, 47).

Type I starch synthases are also highly active in the presence of 0.5 M citrate without the need of a primer (20, 30, 38, 39, 41, 47). Type II enzymes show little or no activity under those conditions. The soluble starch synthases, Type I and II, are immunologically distinct, suggesting that these enzymes are products of the expression of different genes (47). This concept is consistent with a previous finding that the maize endosperm mutant dull is lacking in starch synthase II activity (7).

The starch-bound starch synthase I can also be distinguished from the soluble starch synthase I on the basis of its immunochemical reaction to antibodies prepared against it as compared to the small amount of reaction with the soluble starch synthase I antibody, its different mol wt and, to some extent, on its kinetic properties (47). This also suggests that different genes may be involved in their synthesis.

Similar findings have been observed with the branching enzymes (47). Multiple forms have been observed in extracts of spinach leaf (20) and maize endosperm (5). Three forms have been found in normal maize endosperm designated as I, IIa, and IIb. These forms can be distinguished via kinetic assays (5) and can be distinguished via their immunological reactivity (47). According to Fisher and Boyer (15) antibody prepared against branching enzyme I in rabbit did not react with IIa and IIb, and antibodies prepared toward IIa and IIb did not react with branching enzyme I in enzyme neutralization assays. However, polyclonal antibody prepared in mouse against branching enzyme I or IIa and IIb showed reactivity with all branching enzyme forms in an enzyme-linked immunoabsorbent assay (ELISA). This indicated that the three enzyme forms are indeed related with respect to structure, as well as amino acid composition.

Recently, we have prepared monoclonal antibodies toward branching enzymes IIa plus IIb (47). Some have been shown to react with all three enzymes in the ELISA assay, while others react only with branching enzymes IIa and IIb. These results suggest that all three enzymes are related in structure, but that there are some distinctive differences between branching enzyme I and branching enzymes IIa and IIb. These differences have been observed in the amino acid analysis of the three forms. Some slight differences are seen between branching enzymes IIa and IIb, but they could be within experimental error.

When the purified, essentially homogenous enzymes have been subjected to trypsin and chymotrypsin digestion, and the digests analyzed via HPLC (B. K. Singh and J. Preiss; unpublished results), the retention times and number of peptides of branching enzymes IIa and IIb are very similar. In contrast, the number of peptides, as well as their retention times, seen with the branching enzyme I digest, are quite different. Similar results were obtained with chymotrypsin digests (B. K. Singh and J. Preiss; unpublished results). These data also strongly indicate a difference in structure and sequence between branching enzyme I from IIa and IIb and the very great similarity between IIa and IIb.

Thus, maize endosperm seems to contain at least two types of branching enzyme. They can be distinguished by their binding to DEAE-sepharose, a small difference in mol wt (branching enzyme I mol wt is 82,000, while IIa and IIb are about 80,000), their different reactivities in the two branching enzyme assays, their difference in K_m values for amylose in the I_2 assay, as well as their distinction in the monoclonal antibody reactions.

Distinct multiple forms of branching enzymes and starch synthases have been found in a variety of plant tissues. An obvious question arises. What is the physiological significance for having multiple forms of these two starch biosynthetic enzymes? A reason may be found by considering the structure of the product which these two enzymes are involved in synthesizing, amylopectin. Amylopectin is a very asymmetric type of molecule, and it has been shown that the unit chains of amylopectin fall into two groups; one group of chains having more than 49 glucosyl units long and another group having chain lengths between 12 to 42 glucosyl units (18, 63). Two proposed models of the amylopectin structure by Dexter French (16) are seen in Figure 1. These models, based on various physical and chemical studies of crystalline derivatives of amylopectin, postulate the unit chains being in cluster formations and many of these chains in crystalline structure may be in helical structure. One structure is seen as a modified trichitic structure (I) and the other as a racemose structure (II). Both take into account the variability of the chain size. In order to synthesize these structure types, branching enzymes and starch synthases of different specificity are required for elongation for different chain sizes and for different spacing of the branching points. New assays and primer substrates may have to be devised in order to determine the different specificities of elongation and branching.

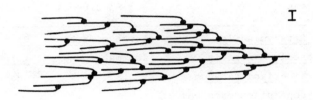

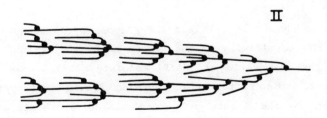

FIG. 1. Proposed models for the structure of amylopectin by French (16). I, A modified trichitic structure. II, A racemose type structure. The symbol · indicates the α-1,6 branch points, while the lines indicate the α-1,4 oligosaccharide chains.

REGULATION OF ADP-GLUCOSE PYROPHOSPHORYLASE

It is well known and widely documented that the plant ADP-glucose pyrophosphorylase is activated by 3-P-glycerate (PGA) and inhibited by inorganic phosphate (Pi) (17, 42-44, 46, 48). These activation and inhibition phenomena have been noted for at least 15 leaf sources (5, 10, 43, 49, 52; Table I), in cyanobacteria (29), green algae (53), and at least seven nonphotosynthetic plant tissues, including endosperm (12), embryo (44, 45), mesocarp (44), root (44), and tuber (57, 59). The activation and inhibition phenomena observed in leaves is essentially the same whether it is a plant undergoing photosynthesis either in C_3, C_4 or CAM (55) pathways.

Table I. Leaf ADP-Glucose Pyrophosphorylases Regulated by 3-Phosphoglycerate and Inorganic Phosphate

C_3 Species	C_4 Species	CAM Species
Spinacea oleracea, Beta vulgaris	Sorghum dochma	Hoya carnosa
Lactuca sativa, Phaseolus vulgaris	Zea mays	Xerosicyos danguyi
Triticum aestivum, Nicotiana tabacum		
Hordeum vulgare, Lycopersicon esculentum		
Oryza sativa, Vitis vinifera		
Manihot esculenta		

It has been shown for all leaf ADP-glucose pyrophosphorylases that PGA modulates the sensitivity of the enzyme to Pi inhibition, and it has been proposed that the ratio of PGA/Pi modulates ADP-glucose and starch synthesis. Thus, this ratio would increase during photosynthesis and carbon fixation in the light, enhancing starch synthesis, and would decrease at night, when synthesis would be inhibited and degradation stimulated. A number of studies either using intact chloroplasts or leaf systems have shown that changes in the PGA/Pi ratio during photosynthesis can be correlated with changes in rate of starch formation (21, 22, 25, 26, 51, 54, 60), thus supporting the above proposal. These various experiments have been previously reviewed (43).

Another factor involved in regulation of starch synthesis may be the ATP concentration. The energy charge value in the chloroplast, the site of starch synthesis, drastically decreases in the dark (27), and the ATP value of 1.0 mM in the light decreases to about 0.2 mM (25). Since the Pi concentration in the dark would also increase, starch degradation would also ensue because of phosphorylase stimulation.

In looking for other metabolites that may affect ADP-glucose synthesis, it was found that inorganic pyrophosphate (PPi) was a potent inhibitor of spinach leaf ADP-glucose pyrophosphorylase (Fig. 2). Fifty per cent inhibition is observed at 0.11 mM when the glucose-1-P concentration is 0.05 mM. If the concentration of glucose-1-P is raised to 0.5 mM, the concentration of pyrophosphate concentration required for 50% inhibition is increased to 0.17 mM. The inhibition appears to be mixed-type inhibition (Fig. 3) and the Ki for PPi, as determined by Dixon Plots, is 60 µM. Figure 4 shows the effect of PPi on the ATP saturation curve. A reciprocal plot shows (Fig. 4B) the inhibition is also mixed type, but there is little effect on the K_m for ATP; it increases from 50 to 70 µM (Table II). Calculation of the Ki for PPi from the data in Figure 4 yields a value of 85 µM, which is similar to the value obtained from the glucose-1-P kinetic data. The effect of PPi on the other kinetic parameters of the spinach leaf enzyme are seen in Table II and in Figure 5. PPi had no effect on the PGA curve nor on the Pi inhibition curve affecting neither the concentration giving 50% activation or inhibition, nor the shape of the activation or inhibition curve. The MgCl$_2$ curve is affected as 0.3 mM pyrophosphate does increase the $S_{0.5}$ (see Table II) value from 1.4 to 2.4 mM.

PPi may thus be a negative modulator of starch synthesis of ADP-glucose synthesis. The major metabolite that may significantly reverse the inhibition would be the substrate, α-glucose-1-P.

Results in our laboratory have shown that the purified potato tuber ADP-glucose pyrophosphorylase is inhibited by PPi, and it has been shown that maize endosperm enzyme is inhibited by PPi (1).

It has been previously thought that PPi formed in various biosynthetic reactions was cleaved by inorganic pyrophosphatase and, thus, was never present in the cell in substantial concentration. Recent results indicated that PPi in animals and bacteria is present in considerable

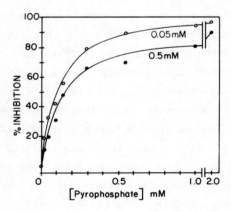

FIG. 2. Inhibition of spinach leaf ADP-glucose pyrophosphorylase by pyrophosphate at the indicated concentrations of α-glucose-1-phosphate. The assay procedure was previously described (10).

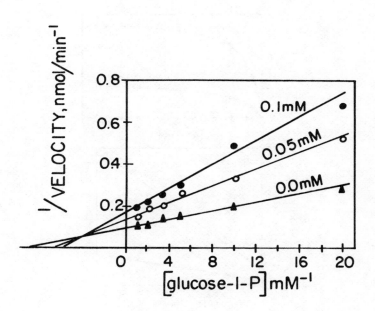

FIG. 3. Inhibition of ADP-glucose synthesis by pyrophosphate. The data are presented as a reciprocal plot of velocity versus glucose-1-phosphate concentration and the pyrophosphate concentrations are indicated as numbers on the lines.

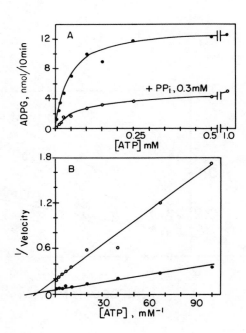

FIG. 4. The effect of pyrophosphate on the ATP saturation curve of spinach leaf ADP-glucose pyrophosphorylase. Figure B is a reciprocal plot of the data presented in Figure A.

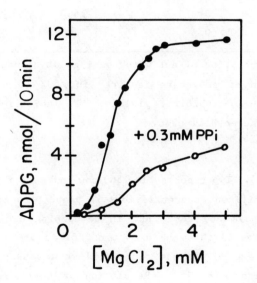

FIG. 5. Effect of pyrophosphate on the MgCl$_2$ saturation curve of spinach leaf ADP-glucose pyrophosphorylase.

Table II. Effect of Pyrophosphate on the Kinetic Parameters of the Spinach Leaf ADP-Glucose Pyrophosphorylase[a]

Substrate/Effector	PPi Concentration (mM)	$A_{0.5}$ (μM)	$S_{0.5}$ (μM)	$I_{0.5}$ (mM)	n	PPi Inhibition (%)
3-P-Glycerate	0.0	38			1.0	0
	0.3	42			1.0	60
ATP	0.0		50		1.0	0
	0.3		74		1.0	56
α-Glucose-1-P	0.0		100		1.0	0
	0.05		140		1.0	38
	0.1		167		1.0	50
Phosphate	0.0				3.9	0
	0.05				3.2	15
	0.1				3.8	49
$MgCl_2$	0		1400		2.6	0
	0.3		2400		2.6	58

[a] $A_{0.5}$, concentration of activator giving 50% of maximal activation; $S_{0.5}$, concentration of substrate required for 50% of maximal velocity; $I_{0.5}$, concentration of inhibitor yielding 50% inhibition; n, Hill constant, slope of line on Hill plot.

concentrations (28). Two reports (14, 56) now indicate that pyrophosphate is present in significant concentrations in plant tissue, being found in shoots, roots of peas and corn, and in developing cotyledons of peas. Indeed, it is estimated that the amount of PPi is at least 14 to 70% of that of ATP (56), and PPi is now recognized as a substrate for a number of reactions in plants (8, 11). Regulation of starch synthesis by PPi could be thought of as a coordination of biosynthetic pathways in the plastid for cell growth with starch synthesis. During growth, the various reactions involved in the synthesis of cellular constituents and macromolecules would be optimal and the levels of PPi could be relatively high since it is a product of the synthetic reactions. The flow of carbon would therefore be for formation of those constituents in the plastid rather than to starch. If synthesis of these constituents is decreased, thus lowering the rate of PPi production, further reduction of PPi concentration would occur by inorganic pyrophosphatase action allowing ADP-glucose and starch synthesis to proceed. Biosynthesis of macromolecules, such as RNA, DNA, and protein, and synthesis of fatty acids occur in the chloroplast and would contribute to PPi formation. PPi could also be formed under some conditions due to photophosphorylation (3). Thus, it would be of interest to determine PPi concentrations in the chloroplast or amyloplast to determine whether it could be a significant factor in regulation of starch synthesis.

In summary, starch synthesis is regulated at the level of ADP-glucose formation by PGA and Pi and possibly by PPi in the plant cells.

Another aspect of regulation of starch synthesis should also be considered since it would also affect carbon flow. During the development of maize endosperm, it has been noted that there is about a 50- to 180-fold increase per kernel in the starch biosynthetic enzyme activities of ADP-glucose pyrophosphorylase and starch synthase (37, 61). Similar increases of the starch biosynthetic enzymes also have been observed in potato tubers (58), in pea fruit (62), and in barley (4). Thus, the genetic regulation of the starch biosynthetic enzymes also plays an integral role for the development of the reserve tissue "sinks."

REGULATION OF STARCH SYNTHESIS IN GUARD CELLS

In contrast to what is observed in the normal mesophyll leaf cell, the guard cells degrade starch during the day, while controlling stomatal opening and closing (34). Starch in the guard cells accumulates at night, once again in contrast to the mesophyll cell. This raised the possibility that regulation of the guard cell ADP-glucose pyrophosphorylase was somewhat different than what was observed for the mesophyll cell enzyme. However, studies with the ADP-glucose pyrophosphorylase isolated from guard cells of Commelina communis (50) and the Argenteum mutant of Pisum sativum (47) show that the guard cell enzyme has the same regulatory properties as the mesophyll leaf cell enzyme (Table III). As seen in

Table III, the guard cell preparations appear to be quite free of mesophyll cell contamination since the ribulose-1,5-bisphosphate carboxylase (Rubisco) activity is very low. PGA activates ADP-glucose synthesis and Pi inhibits it. It has also been shown for the C. communis enzyme that PGA overcomes Pi inhibition (50) and that increasing the concentration of Pi reverses PGA activation (47), a phenomenon observed many times with the mesophyll cell ADP-glucose pyrophosphorylases. It has also been shown that ADP-glucose pyrophosphorylase in the Vicia fabia guard cell is inhibited by Pi and activated by PGA (36).

Table III. Effect of Phosphoglycerate and Phosphate on ADP-Glucose Pyrophosphorylase Activity of Commelina communis and Pisum sativum Mesophyll and Guard Cells

Plant	Activity/Conditions	Guard Cells	Mesophyll Cells
		nmol min^{-1} mg^{-1}	
Commelina communis	ADP-glucose pyrophosphorylase, complete	31	92
	-PGA, 1 mM	3.6	
	+Pi, 1 mM	3.1	
	Ribulose-1,5-bis-P carboxylase	4.2	240
Pisum sativum	ADP-glucose pyrophosphorylase, complete	12	72
	-PGA, 1 mM	< 0.2	
	+Pi, 1 mM	< 0.2	
	Ribulose-1,5-bis-P carboxylase	12	340

Because the guard cell appears not to contain Rubisco activity (33, 35) and, therefore, cannot form PGA during photosynthesis or at night from CO_2, it would be of interest to know the source of PGA, since it is an almost absolute requirement for activation of the ADP-glucose pyrophosphorylase.

It is well known that the mesophyll cells of C_3 plants have starch and its biosynthetic enzymes are localized in the chloroplast (32). If this is also true for the guard cell, then the source of PGA and triose-phosphates (triose-P) for the guard cell chloroplast must be derived from the cytosol and possibly from other cell types. If the chloroplast in the guard cell is similar to that of the mesophyll cell, the only glycolytic metabolites appreciably permeable to the guard cell chloroplast would be the triose-P and PGA. The transport of the triose-P into the chloroplast would occur by exchange with Pi via the triose-P/Pi translocator. It may be possible that a carbon source such as sucrose is transported from mesophyll cells into the guard cell. For entrance of this carbon source into the chloroplast, it would first have to be metabolized to the triose-P/PGA level. The ATP required for starch synthesis in the guard cell could arise from oxidative phosphorylation with malate as the electron source, since malate accumulates in the guard cell from starch breakdown during the day.

In order to obtain more information on starch metabolism in the guard cell, we decided to determine if the starch biosynthetic enzymes were indeed localized in the guard cell chloroplast. Thus, the abaxial epidermis of C. communis leaves were peeled and digested with 2% cellulase at first in a hypotonic solution containing 0.25 M mannitol and then with 2% cellulase and 0.5% macerase in 0.4 M mannitol to prepare guard cell protoplasts. The protoplasts were then isolated from a discontinuous percoll gradient of 22, 67, and 90% percoll. They were then lysed by passage through a 5-micron mesh in a microfuge. The chloroplast pellet was resuspended and chloroplast lysed with a Pasteur pipet in hypotonic solution.

Table IV shows that 90% of the PEP carboxylase activity is found in the supernatant fraction as expected, while 80% of the starch biosynthetic enzyme activities are mainly associated with the chloroplast fraction. Cytochrome c oxidase activity was found mainly in the supernatant fraction,

Table IV. Subcellular Localization of the Starch Biosynthetic and Degradative Enzymes of Commelina communis Guard Cells

	Units of Activity Found In:[a]		
	Supernatant	Chloroplast	Average Per Cent in Chloroplast
PEP carboxylase	13.0	1.89	9.8 $\pm$ 4.9 (17)
ADPG pyrophosphorylase	2.1	8.0	80.1 $\pm$ 5.7 (22)
Starch synthase	0.12	0.88	83.0 $\pm$ 7.1 (2)
Branching enzyme	20	65	75.0 $\pm$ 8.2 (3)
Amylase	5.3	2.1	28.8 $\pm$ 6.7 (5)
Phosphorylase	3.7	2.2	42.7 $\pm$ 6.3 (4)
Debranching (R) enzyme	3.75	1.3	29.2 $\pm$ 3.9 (4)
Cytochrome c oxidase	1.68	0.37	18.0 $\pm$ 0.0 (2)

[a] Units are expressed as nmol of product formed per min. The activities shown are only representative values. The average per cent and the standard deviation found in the chloroplast were determined from the number of experiments indicated in parentheses. The starch biosynthetic and degradative enzymes were assayed as described previously (32). The assay for PEP carboxylase was done essentially in the same manner as the Rubisco assay, except that P-enolpyruvate was the substrate (2) and the cytochrome c oxidase assay (23) was done as described. ADPG, ADP-glucose.

but 18% of the activity was associated with the chloroplast, indicating some contamination of the chloroplast fraction with mitochondria. Nevertheless, the data indicate that the guard cell starch biosynthetic enzyme activities are, for the major part, associated with the chloroplast, if not entirely present in the chloroplast.

In contrast, the starch degradative enzymes appear to be associated mainly with the supernatant fraction. However, there also appears to be sufficient amylase, phosphorylase, and debranching activities to degrade the synthesized starch in the chloroplast. This conclusion is based on the observation that the degradative activities are at higher levels of activity than the starch synthase activity. However, the observation that the greater portion of the degradative activities reside in the cytosol is surprising. The site of synthesis of starch is considered to be in the chloroplast; finding the biosynthetic enzymes primarily present in the chloroplast fraction is consistent with this view. A function for the starch degradative enzymes in the cytosol remains unknown. However, the finding that a major portion of the starch degradative activities are found in the guard cell cytosol and not in the chloroplast is essentially the same as what has been found for the spinach leaf mesophyll cell (32). Thus, the pattern of distribution of the starch biosynthetic and degradative enzymes seen in the guard cell protoplast is very similar to what has been observed in the protoplasts of spinach leaf mesophyll cells.

The availability of what appear to be intact guard cell chloroplasts may aid in attempting to determine the route of starch biosynthesis in the guard cell, and efforts in these directions are now being made.

THE PRESENCE OF THE STARCH BIOSYNTHETIC ENZYMES IN MAIZE LEAF BUNDLE SHEATH AND MESOPHYLL CELLS

The major portion of starch formation in a C_4 plant leaf occurs generally in the bundle sheath cell; the site of PGA formation via the Rubisco enzyme-catalyzed reaction. However, there are certain growth conditions where the maize leaf mesophyll cell would accumulate significant amounts of starch. It has been reported that this can occur when the maize leaf is subjected to continuous light for at least 2 d (13). It was of interest for us to determine the enzyme levels of the starch biosynthetic enzymes. Although there are some reports on the localization of either starch synthase, ADP-glucose pyrophosphorylase, or phosphorylase in some C_4 plants (9, 13, 24, 31), the determination of branching enzyme activity has not been done. Nor has any mention been made of the properties of the ADP-glucose pyrophosphorylase with respect to its activation by PGA. In some cases, the purity of the cell type being studied was not verified.

TABLE V. Starch Biosynthetic Enzyme Levels in Maize Leaf Bundle Sheath and Mesophyll Cells

	Leaf Extract	Bundle Sheath	Mesophyll
	nmol min^{-1} mg^{-1} protein		
Rubisco	140	150	0.0
PEP carboxylase	660	6	960
ADPG pyrophosphorylase	70	220	12
Starch synthase	22	20	14
Branching enzyme	138	505	373
Phosphorylase	27	17	30

Table V shows the levels of the starch biosynthetic enzymes and phosphorylase in the maize leaf bundle sheath and mesophyll cells. Rubisco and PEP carboxylase assays were also done to determine the possible contamination of the cell types. The preparations were judged to be over 99% pure on the basis of the Rubisco and PEP carboxylase assays. As one would guess, the levels of the starch biosynthetic enzymes are higher in the bundle sheath cell, particularly the ADP-glucose pyrophosphorylase. In contrast, phosphorylase which is considered a degradative enzyme appears to predominate in the mesophyll cell. This result is similar to that obtained by Mbaku et al. (31), where phosphorylase was found mainly in the mesophyll cell of Digitaria pentzii. It is possible that the balance between starch biosynthetic activity levels and starch degradative activity levels may play a part in determining the levels of starch in the different cells. Other experiments directed toward determining amylase and debranching (R) enzyme activities will also be done to determine whether their distribution would be similar to phosphorylase.

The ability of PGA to stimulate ADP-glucose pyrophosphorylase activity in the crude extracts of bundle sheath and mesophyll cells was tested, and the results are seen in Figure 6. Both enzymes are activated by PGA. However, the bundle sheath enzymes can be stimulated over 20-fold with 1.0 mM PGA, while the mesophyll ADP-glucose pyrophosphorylase is only

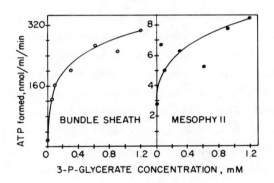

FIG. 6. Activation of the <u>Zea</u> <u>mays</u> bundle sheath and mesophyll cell ADP-glucose pyrophosphorylase by 3-phosphoglycerate. The assay (10) was done in the pyrophosphorylase direction. The reaction mixture contained glycylglycine buffer, pH 7.3, 6 mM $MgCl_2$, 1 mM ADP-glucose, 1 mM 3-P-glycerate, and 2 mM ^{32}PPi.

stimulated 3-fold. Moreover, the unactivated activity observed in the bundle sheath is greater than the activated activity of the mesophyll cell. Thus, it is likely that greater starch synthesis occurs in the bundle sheath because of the higher enzyme activity, as well as the greater activation of the bundle sheath ADP-glucose pyrophosphorylase by PGA. The enzyme from both cell types is not activated by pyruvate and not to a great extent by PEP. Phosphate may also play a role in the regulation of starch synthesis and, thus, further studies of the ADP-glucose pyrophosphorylase in the two cell types would be done to determine whether there are differences in the sensitivities toward phosphate inhibition.

As suggested by Downton and Hawker (13), the mesophyll cell may be more involved in sucrose synthesis than in starch synthesis. If so, then the lesser sensitivity of the mesophyll ADP-glucose pyrophosphorylase to PGA activation may be rationalized as a mechanism to avoid or minimize the competition for carbon flow toward sucrose formation.

In summary, synthesis of starch entails the use of a number of multiple forms of enzymes required to form a branched product that is asymmetric in structure. The nature of the regulation of the synthesis of the glucosyl donor for starch synthesis appears to be a constant feature whether the cell synthesizing starch is obtaining its carbon either from photosynthetic activity or from nonphotosynthetic activity. It is quite possible that the reason for this is that synthesis of starch occurs in a plastid in either photosynthetic or nonphotosynthetic tissue and that the entry of carbon into the plastid may predominantly be via the triose-P/phosphate translocator and, thus, involve movement of those compounds which are the allosteric effectors of the enzyme catalyzing synthesis of ADP-glucose, the glucosyl donor for starch synthesis.

Acknowledgments--The research described herein was supported by NSF Grant PCM 82-05705 and a McKnight grant to N.R.

LITERATURE CITED

1. AMIR J, JH CHERRY 1972 Purification and properties of adenosine diphosphoglucose pyrophosphorylase from sweet corn. Plant Physiol 49:893-897

2. BAHR JT, PG JENSEN 1978 Activation of ribulose bisphosphate carboxylase in intact chloroplasts by CO_2 and light. Arch Biochem Biophys 185:39-48
3. BALTSCHEFSKY H, LV VON STEDINGK, HW HELDT, M KLINGENBERG 1966 Inorganic pyrophosphate: formation in bacterial photophosphorylate. Science 153:1120-1122
4. BAXTER ED, CM DUFFUS 1971 Starch synthetase in developing barley amyloplasts. Phytochemistry 10:2641-2644
5. BOYER CD, J PREISS 1978 Multiple forms of (1-4)-α-D-glucan (1-4)-α-D-glucan 6 glycosyl transferase from developing Zea mays L. kernels. Carbohydr Res 61:321-334
6. BOYER CD, J PREISS 1979 Properties of citrate-stimulated starch synthesis catalyzed by starch synthase I of developing maize kernels. Plant Physiol 64:1039-1042
7. BOYER CD, J PREISS 1981 Evidence for independent genetic control of the multiple forms of maize endosperm branching enzymes and starch synthases. Plant Physiol 67:1141-1145
8. CARNAL NW, CC BLACK 1979 Pyrophosphate-dependent 6-phosphofructokinase a new glycolytic enzyme in pineapple leaves. Biochem Biophys Res Commun 86:20-26
9. CHEN TM, P DITTRICH, WH CAMPBELL, CC BLACK 1974 Metabolism of epidermal tissues, mesophyll cells and bundle sheath strands resolved from mature nutsedge leaves. Arch Biochem Biophys 163:246-262
10. COPELAND L, J PREISS 1981 Purification of the spinach leaf ADPglucose pyrophosphorylase. Plant Physiol 68:996-1001
11. CSEKE C, NF WEEDEN, BB BUCHANAN, K UYEDA 1982 A special fructose bisphosphate functions as a cytoplasmic regulatory metabolite in green leaves. Proc Natl Acad Sci USA 79:4322-4326
12. DICKINSON D, J PREISS 1969 ADPglucose pyrophosphorylase from maize endosperm. Arch Biochem Biophys 130:119-128
13. DOWNTON WJS, JS HAWKER 1973 Enzyme of starch and sucrose metabolism in Zea mays leaves. Phytochemistry 12:1551-1556
14. EDWARDS J, T AP REES, PM WILSON, S MORREL 1985 Measurement of the inorganic pyrophosphate in tissues of Pisum sativum L. Planta. 162:188-191

15. FISHER MB, CD BOYER 1983 Immunological characterization of maize starch branching enzymes. Plant Physiol 72:813-816
16. FRENCH D 1972 Fine structure of starch and its relationship to the organization of starch granules. J Jap Soc Starch Sci 19:8-25
17. GHOSH HP, J PREISS 1966 Adenosine diphosphate glucose pyrophosphorylase - a regulatory enzyme in the biosynthesis of starch in spinach leaf chloroplasts. J Biol Chem 241:4491-4504
18. GUNJA-SMITH Z, JJ MARSHALL, C MERCIER, EE SMITH, WJ WHELAN 1970 A revision of the Meyer-Bernfield model of glycogen and amylopectin. FEBS Lett 12:101-104
19. HAWKER JS, WJS DOWNTON 1974 Starch synthesis from *Vitis vinifera* and *Zea mays*. Phytochemistry 13:893-900
20. HAWKER JS, JL OZBUN, H OZAKI, E GREENBERG, J PREISS 1974 Interaction of spinach leaf adenosine diphosphate glucose α-1,4 glucan α-4 glucosyl transferase and α-1,4 glucan, α-1,4 glucan-6-glycosyl transferase in synthesis of branched α-glucan. Arch Biochem Biophys 160:530-551
21. HELDT HW, CJ CHON, D MARONDE, A HEROLD, ZS STANKOVIC, DA WALKER, A KRAMINER, MR KIRK, U HEBER 1977 Role of orthophosphate and other factors in the regulation of starch formation in leaves and isolated chloroplasts. Plant Physiol 59:1146-1155
22. HEROLD A, DH LEWIS, DA WALKER 1976 Sequestration of cytoplasmic orthophosphate by mannose and its differential effect on photosynthetic starch synthesis in C3 and C4 species. New Phytol 76:397-407
23. HODGES TK, RT LEONARD 1974 Purification of a plasma membrane bound adenosine triphosphatase from plant roots. Methods Enzymol 32:392-406
24. HUBER W, MAR DeFEKETE, H ZIEGLER 1969 Enzyme des starke-umsatzes in bundelscheiden- und palisadenchloroplasten von *Zea mays*. Planta 87:360-364
25. KAISER WM, JA BASSHAM 1979 Light-dark regulation of starch metabolism in chloroplasts I. Levels of metabolism in chloroplasts and medium during light-dark transition. Plant Physiol 63:105-108

26. KAISER WM, JA BASSHAM 1979 Light-dark regulation of starch metabolism in chloroplasts II. Effect of chloroplast metabolite levels on the formation of ADPglucose by chloroplast extracts. Plant Physiol 63:109-113
27. KOBAYASHI Y, Y INOUE, F FURUYA, K SHIBATA, U HEBER 1979 Regulation of adenylate levels in intact spinach chloroplasts. Planta 147: 69-75
28. KUKKO E, J HEINONEN 1982 The intracellular concentration of pyrophosphate in the batch culture of Escherchia coli. Eur J Biochem 127: 347-349
29. LEVI C, J PREISS 1976 Regulatory properties of the ADPglucose pyrophosphorylase of the blue green bacterium, Synechococcus 6301. Plant Physiol 58:753-756
30. MacDONALD FD, J PREISS 1983 Solubilization of the starch-granule-bound starch synthase of normal maize kernels. Plant Physiol 73: 175-178
31. MBAKU SB, GJ FRITZ, G BOWES 1978 Photosynthetic and carbohydrate metabolism in isolated leaf cells of Digitaria pentzii. Plant Physiol 62:510-515
32. OKITA TW, E GREENBERG, DN KUHN, J PREISS 1979 Subcellular localization of the starch degradative and biosynthetic enzymes of spinach leaves. Plant Physiol 64:187-192
33. OUTLAW WH 1982 Carbon metabolism in guard cells. Rec Adv Phytochem 16:185-222
34. OUTLAW WH, J MANCHESTER 1979 Guard cell starch concentration quantitatively related to stomatal aperture. Plant Physiol 64:79-82
35. OUTLAW WH, J MANCHESTER, CA DICOMELLI, DD RANDALL, B RAPP, GM VIETH 1979 Photosynthetic carbon reduction pathway is absent in chloroplasts of Vicia fabia guard cells. Proc Natl Acad Sci USA 76: 6371-6375
36. OUTLAW WH, MC TARCZYNSKI 1984 Guard cell starch biosynthesis regulated by effectors of ADPglucose pyrophosphorylase. Plant Physiol 74:424-429

37. OZBUN JL, JS HAWKER, G GREENBERG, C LAMMEL, J PREISS, EYC LEE 1973 Starch synthase, phosphorylase, ADPglucose pyrophosphorylase, and UDPglucose pyrophosphorylase in developing maize kernels. Plant Physiol 51:1-5
38. OZBUN JL, JS HAWKER, J PREISS 1971 Adenosine diphosphoglucose-starch glucosyl transferases from developing kernels of waxy maize. Plant Physiol 48:765-769
39. OZBUN JL, JS HAWKER, J PREISS 1972 Soluble adenosine diphosphate glucose-α-1,4 glucan α-4-glucosyl-transferases from spinach leaves. Biochem J 126:953-963
40. PISIGAN RA, EJ DEL ROSARIO 1976 Isoenzymes of soluble starch synthetase from Oryza sativa grains. Phytochemistry 15:71-73
41. POLLOCK C, J PREISS 1980 The citrate stimulated starch synthase of starch maize kernels: purification and properties. Arch Biochem Biophys 204:578-588
42. PREISS J 1982 Biosynthesis of starch and its regulation. In FA Loewus and W Tanner, eds, Plant Carbohydrates, Encyclopedia of Plant Physiology, Vol 13A, New Series. Springer-Verlag, Berlin, Heidelberg, pp 397-417
43. PREISS J 1982 Regulation of the biosynthesis and degradation of starch. Annu Rev Plant Physiol 33:431-454
44. PREISS J, HP GHOSH, J WITTKOP 1967 Regulation of the biosynthesis of starch in spinach leaf chloroplasts. In TW Goodwin, ed, Biochemistry of Chloroplasts, Vol II. Academic Press, New York, pp 131-153
45. PREISS J, C LAMMEL, A SABRAW 1971 A unique adenosine diphosphoglucose pyrophosphorylase associated with maize embryo tissue. Plant Physiol 47:104-108
46. PREISS J, C LEVIS 1980 Starch biosynthesis and degradation. In J Preiss, ed, Biochemistry of Plants, Vol 3. Academic Press, New York, pp 371-423
47. PREISS J, FD MacDONALD, BK SINGH, N ROBINSON, K McNAMARA 1985 Various aspects in the regulation of starch biosynthesis. In L Munck and R Hill, eds, Proc Int Conf on New Approaches to Research on Cereal Carbohydrates. Elsevier Science Publishers, Amsterdam, pp 1-11

48. PREISS J, DA WALSH 1981 The comparative biochemistry of glycogen and starch. In V Ginsburg, ed, Biology of Carbohydrates, Vol 1. J Wiley Press, New York, pp 199-314
49. RIBEREAU-GAYON G, J PREISS 1971 ADPglucose pyrophosphorylase from spinach leaf. Methods Enzymol 23:618-624
50. ROBINSON NL, E ZEIGER, J PREISS 1983 Regulation of ADPglucose synthethis in guard cells of Commelina communis. Plant Physiol 73:862-864
51. SANTARIUS KA, U HEBER 1965 Changes in the intracellular levels of ATP, ADP, AMP, and Pi and regulatory function of the adenylate system in leaf cells during photosynthesis. Biochim Biophys Acta 102:39-54
52. SANWAL GG, E GREENBERG, J HARDIE, EC CAMERON, J PREISS 1968 Regulation of starch biosynthesis in plant leaves: activation and inhibition of ADPglucose pyrophosphorylase. Plant Physiol 43:417-427
53. SANWAL GG, J PREISS 1967 Biosynthesis of starch in Chlorella pyrenoidosa II regulation of ATP: α glucose-1-phosphate adenylyl transferase (ADPglucose pyrophosphorylase) by inorganic phosphate and 3-phosphoglycerate. Arch Biochem Biophys 119:454-469
54. SHEU-HWA CS, DH LEWIS, DA WALKER 1975 Stimulation of photosynthetic starch formation by sequestration of cytoplasmic orthophosphate. New Phytol 74:383-392
55. SINGH BK, E GREENBERG, J PREISS 1984 ADPglucose pyrophosphorylase from the CAM plant Hoya carnosa and Xeroxicyos danguyi. Plant Physiol 74:711-716
56. SMYTH DA, CC BLACK 1984 Measurement of the pyrophosphate content of plant tissues. Plant Physiol 75:862-864
57. SOWOKINOS JR 1981 Pyrophosphorylases in Solanum tuberosum II. Catalytic properties and regulation of ADPglucose and UDPglucose pyrophosphorylase activities in potatoes. Plant Physiol 68:924-929
58. SOWOKINOS JR 1976 Pyrophosphorylases in Solanum tuberosum I. Changes in ADPglucose and UDPglucose pyrophosphorylase activities associated with starch biosynthesis during tuberization, maturation, and storage of potatoes. Plant Physiol 57:63-68

59. SOWOKINOS JR, J PREISS 1982 Pyrophosphorylases in *Solanum* *tuberosum* III Purification and regulatory properties of potato tuber ADP-glucose pyrophosphorylase. Plant Physiol 69:1459-1466
60. STEUP M, DG PEAVY, M GIBBS 1976 The regulation of starch metabolism by inorganic phosphate. Biochem Biophys Res Commun 72:1554-1561
61. TSAI CY, F SALAMINI, OE NELSON 1970 Enzymes of carbohydrate metabolism in the developing endosperm of maize. Plant Physiol 46:299-306
62. TURNER JF 1969 Physiology of pea fruits VI. Changes in uridine diphosphate glucose pyrophosphorylase and adenosine diphosphate glucose pyrophosphorylase in the developing seed. Aust J Biol Sci 22:1145-1151
63. WHELAN WJ 1971 Enzymatic explorations of the structures of starch and glycogen. Biochem J 122:609-622

The Degradation of Transitory Starch Granules in Chloroplasts

ERWIN BECK

Depending on the species and the environmental conditions, green plants can deposit a major portion of their photosynthetic carbon gain as starch or starch-like polysaccharides within the chloroplast. Figure 1A shows typical granules of assimilatory starch in a spinach chloroplast. These starch granules disappear more or less completely during the nocturnal dark period (Fig. 1B) and therefore have been also named transitory starch. The biosynthetic route and the enzymology of the synthesis of chloroplastic starch have been quite well characterized (18, 19). However, considerable controversy concerning the mode of the degradation of this starch within the chloroplast is found in the literature, whereby the main discussion focuses on the question of whether hydrolysis by amylases, debranching enzyme and D-enzyme, or the energetically favorable phosphorolysis by phosphorylase represents the predominant pathway. Since carbon export via triose-phosphates or PGA is much more rapid than that via glucose (5) and since phosphorylase has been observed in all chloroplast preparations examined, the verdict in favor of phosphorolysis has been pronounced by many researchers (e.g. 11, 28, 29, 30); yet the formation of maltose and glucose has also been repeatedly reported when starch breakdown was studied with isolated chloroplasts or protoplasts (7, 10, 16). The arguments against hydrolysis as the major type of reaction in chloroplastic starch degradation are that glucose--which can permeate the envelope (21)--represents only a by-product originating from either maltose phosphorylase (22) or the glucan:glucose glucosyl-transferase reaction (12), and that the capacity of the chloroplast membranes for maltose transport is very limited (7). On the other hand, amylolytic activities have been observed in the chloroplasts of several species which were equivalent to, or even in excess of, that of phosphorylase, when both types of enzymic reactions were assayed under optimal conditions (4, 17). In order to be able to comment on the respective importance of amylolysis and phosphorolysis, the following questions must be posed: which types of hydrolytic activities are present in the chloroplast, how are they regulated, and can the assimilatory starch granule serve as a substrate for both amylases and phosphorylase?

With respect to the last question, more data on the structure and composition of the chloroplastic starch granule are required and, therefore, the first part of this paper is devoted to that subject.

Different hydrolytic activities acting on starch can be identified by electrophoresis in polyacrylamide gels containing different species of starch such as amylose, β-limit dextrin, and amylopectin (2, 8). In order to study the regulation of the starch-degrading activities, the various enzymes must then be purified and characterized. Up to the present, this has been achieved with an α-amylase (15), with R-enzyme (13, 15), and with phosphorylase (23, 27) from spinach leaves.

Finally, the question will be dealt with as to whether the chloroplast envelope provides a special translocating system for maltose and, if so, whether phosphorolytic starch breakdown is actually capable of preserving more energy than amylolytic degradation.

RESULTS AND DISCUSSION

<u>The Chloroplastic Starch Granule</u>. The data reported here were obtained with spinach and have been confirmed at least partly by studies with pea leaves. Assimilatory starch granules were consistently found to be surrounded by the thylakoid system of the plastid. On account of their diurnal growth and degradation, the quotation of an average size is not very meaningful; however, kernels of up to 3 µm in diameter were regularly observed at the end of the daily period of illumination. Chloroplastic starch granules have been prepared from spinach by Steup <u>et al</u>. (24). Upon investigation by scanning electron microscopy, they exhibited sharp contours. When assimilatory starch granules were prepared according

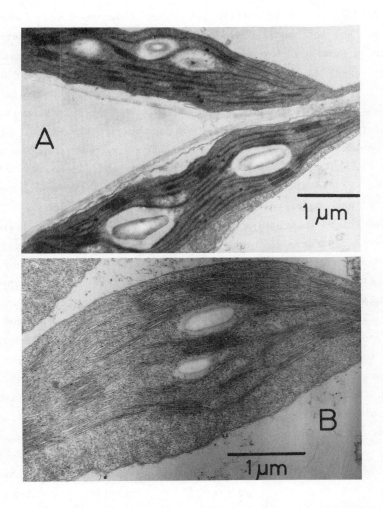

FIG. 1. A, Spinach chloroplasts at the end of the daily period of illumination (12 h) containing large granules of assimilatory starch. (Magnification of X 19,000). Leaf pieces (1 mm^3) were fixed in 3% glutaraldehyde (dissolved in 0.05 M phosphate buffer pH 6.9) for 12 h and embedded in Epon 812. The ultrathin sections were contrasted for 12 min with uranyl-acetate and subsequently for 7 min with Pb(CH$_3$COO)$_2$ before investigation by transmission electron microscopy. B, Spinach chloroplasts at the end of the nocturnal dark period (12 h) showing largely degraded starch granules (X 23,700). Preparation as in A.

to the procedure of these authors (24) from illuminated and from darkened spinach leaves, a dramatic difference in the appearance was observed with the scanning electron microscope. Figure 2A, showing the preparation from darkened leaves, is in accordance with the picture provided by Steup et al. (24). However, assimilatory starch granules from an illuminated leaf appear pasty and lack sharp contours (Fig. 2B). When this preparation was incubated for 6 h with a phosphate-free soluble protein extract of isolated chloroplasts, the pasty cover disappeared and the granules appeared as shown in Figure 2A. The results indicate that the endogenous hydrolytic activities are capable of degrading the pasty layer. However, due to the omission of phosphate, only amylolysis was allowed to proceed.

In addition to scanning electron microscopy, starch granules were investigated by transmission electron microscopy, subsequent to fixation of small leaf pieces with 3% glutaraldehyde. As shown in Figure 1A, a distinct granule core is surrounded by a sizable amount of apparently empty space which could not be contrasted with osmium tetroxide or uranylacetate. In order to investigate whether this area also contains starch, the leaf sections were stained with iodine prior to electron microscopy. Although iodine evaporated rapidly due to the vacuum in the electron microscope, some staining of the light area could be demonstrated (Fig. 3), indicating that the outer layer of the granule also consists of starch. Since native starch has a strong tendency to crystallize (20), it is assumed that the assimilatory starch granule consists of a crystalline core which is surrounded by a noncyrstalline, pasty layer. Several types of starch or glycogen have been distinguished by the absorption spectra of their iodine complexes (9). Generally, branching of the homoglucan results in a shift of the absorption maximum towards shorter wavelengths (Table I). Although it is not possible to separate the two components of the chloroplastic starch granule, differing chemical structures of the core and the mantle are suggested by the following finding: during the nocturnal breakdown of the granule, the mantle appears to be degraded more rapidly than the core (Fig. 4B), resulting in an enrichment of the starch type characteristic of the core. Simultaneously, the maximum of the absorption spectrum of the iodine complex, which is typical of amylopectin, shifted towards longer wavelengths, thus indicating a

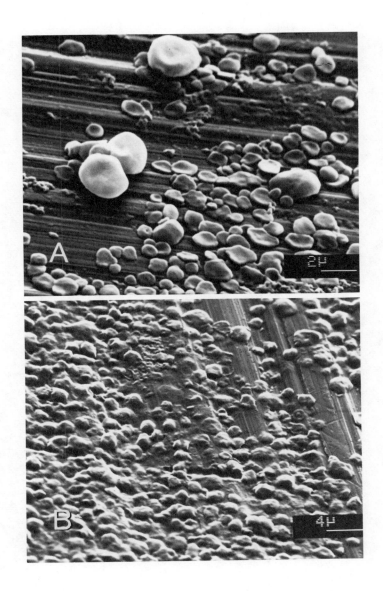

FIG. 2. Scanning electron micrographs of assimilatory starch isolated from darkened (A, top) and from illuminated (B, bottom) spinach leaves. The procedures were described by Steup et al. (2).

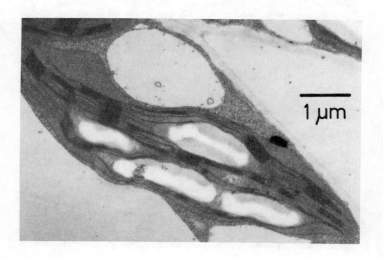

FIG. 3. Assimilatory starch granules in spinach chloroplasts after staining of the ultrathin sections with iodine dissolved in aqueous solution of KI. Other conditions as described for Figure 1A.

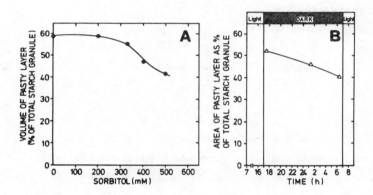

FIG. 4 A, Volume change of the pasty mantle of spinach assimilatory starch granules as caused by increasing osmolarity of sorbitol in the medium used for the fixation of the leaf pieces. The areas of the starch granule and of the pasty portion were determined on the electron micrographs with a planimeter. Differences were statistically significant. B, Preferential degradation of the pasty layer of spinach assimilatory starch during the nocturnal dark period. Preparation of the material as described for Figure 1A. Determination of the areas of the pasty layer and of the core was as in Figure 4A.

Table I

Absorption maxima of the iodine complexes of different kinds of starch and glycogen. The polysaccharides were dissolved at 100°C in a saturated aqueous solution of $CaCl_2$. After cooling, 1 volume of this solution was added to 1 volume of iodine reagent, and the absorption was recorded <u>versus</u> a blank of saturated $CaCl_2$ solution: iodine reagent = 1:1 (v/v). Composition of iodine reagent: 0.1 ml of a solution of 0.26 g of I_2 and 2.6 g of KI in 10 ml H_2O was mixed with 15 ml saturated aqueous $CaCl_2$ solution (9).

Substrate	E_{max} of glucan-iodine complex (nm)
Amylose	645
Spinach assimil. starch (night)	547
Amylopectin	530
Spinach assimil. starch (day)	528
Soluble starch	525
Glycogen	450

decrease in α-1→6 branches (Table I). This finding suggests a higher degree of chain branching in the mantle than in the core.

The idea of a crystalline core and a pasty mantle implies differing water contents in the two parts of the starch granule. In order to test this hypothesis, sections of a spinach leaf were fixed with 3% glutaraldehyde in solutions of increasing osmolarity achieved with sorbitol. From the resulting electron micrographs, the areas of the mantle and the core were determined and plotted <u>versus</u> the osmotic concentration of the fixation medium. Figure 4A demonstrates significantly greater shrinkage of the mantle as compared to that of the core at hypertonic sorbitol concentrations and, thus, indicates that the pasty layer indeed possesses a higher water content, which can be osmotically influenced.

<u>Chloroplastic Amylases and Their Role in Starch Breakdown.</u> Beck <u>et al</u>. (2) have shown that even under optimal conditions, assimilatory starch is a very poor substrate for the chloroplastic phosphorylase in spinach.

However, amylolysis of this polysaccharide resulted in the formation of degradation products that were acted on by phosphorylase three to four times more rapidly than was the case with the starting material. In the light of more recent data on the substrate specificity of spinach chloroplast phosphorylase (25), the low rates of phosphorolytic degradation observed by Beck et al. (2) without prior amylolysis are to be understood in terms of a concomitant amylolytic hydrolysis rather than an expression of the sole action of phosphorylase on the polysaccharide: chloroplastic phosphorylase is capable of cleaving $\alpha-(1\rightarrow 4)$ bonds of only oligoglucans.

These results demonstrate the cooperativity of the amylolytic system and phosphorylase in assimilatory starch breakdown, rather than a competition of the two modes of degradation.

Electrophoresis of chloroplast protein extracts on polyacrylamide gels (containing either amylose, amylopectin or β-limit dextrin for enzymic substrates) and subsequent negative activity staining with iodine revealed endo- and exo-amylases (which correspond to the α- and β-amylases of cereal grains) as well as the debranching or R-enzyme as constituents of the spinach chloroplast (2). On the other hand, pea chloroplasts contained only endo-amylase and R-enzyme (4). These activities cannot be ascribed to impurities of the chloroplast preparations, because gel electrophoresis of a protein extract from whole leaves yielded additional amylase bands in both cases. Maltase activity could not be demonstrated in chloroplast extracts (18).

The total amylolytic activity (including R-enzyme), as determined with a crude protein extract of spinach chloroplasts, exhibited a pH-optimum at 6.1 (17). Thus, under the diurnal pH-regime of the chloroplast stroma (6) only sub-optimal amylolytic activities can be expected in spinach chloroplasts, these being considerably lower during the day (pH 8) than during the night (pH 7). In addition to the regulatory influence of the stromal pH, an endogenous diurnal oscillation of amylolytic activity has been observed with spinach chloroplasts (17), and with cotton leaves (3), resulting in a nocturnal activity (50-100 nmol min^{-1} [mU] mg^{-1} Chl; maltose from polyglucan) approximately double that found during the day (25-30 mU mg^{-1} Chl) in the case of the spinach system.

By superimposing the pH-effect on the diurnal oscillation of the amylolytic system, a minimum activity of 8 to 10 mU mg^{-1} Chl can be assumed for the day, and a maximal activity of about 40 to 50 mU mg^{-1} Chl at the end of the night. The starch content of illuminated spinach chloroplasts was found to be in the range of 50 to 90 μatoms C mg^{-1} Chl (17, 30), which is equivalent to 8.3 to 15 μmol of glycosidic bonds per mg Chl. The nocturnal activity of 40 to 50 mU mg^{-1} Chl would then be sufficient for complete degradation of the polysaccharide to maltose within 2.5 h. However, the amylolytic activity may be counteracted by the glucan: glucose glycosyl transferase (12), which is capable of producing longer-chain maltodextrins from maltose and thereby increases the number of glycosidic bonds accessible to amylases. Yet even with transglycosidase, the amylolytic system of the chloroplast would be easily sufficient for complete degradation of the chloroplastic starch granule to maltose.

Although the starch granule itself is only accessible to the amylolytic system, competition of amylases and phosphorylase for the final degradation of the oligo-maltodextrins must be assumed, and the more favorable the conditions are for phosphorylase (e.g. high concentration of Pi) the higher the yield of glucose-1-P. With starch granules from cotton leaves, the levels of the amylolytic degradation products by far exceeded those of phosphorolysis (3). Starch breakdown in spinach chloroplasts under simulated in vivo conditions resulted in the formation of both types of end products, glucose-1-P and maltose plus glucose (7, 10, 30). On the contrary, no significant amounts of unphosphorylated products have been found with pea chloroplasts (11, 30).

From spinach leaves, two phosphorylases (23, 27) and the debranching enzyme (13) have been purified to homogeneity by affinity chromatography. In addition, an endo-amylase has been studied (15), which, as is the case with one of the phosphorylases (26), is of extrachloroplastic origin. It is of particular interest that the chloroplastic phosphorylase preferentially used unbranched maltodextrins of low mol wt (degree of polymerization, 5-11) as substrates, but was ineffective with branched forms of starch, whereas the cytosolic isozyme was more effective with branched glucans of high mol wt (25). Thus, spinach chloroplast phosphorylase is only capable of participating in the final degradation of the products of amylolysis, and cannot initiate the breakdown of the granule. Chloroplastic endo-amylase may be substantially associated with the starch granules as long as these exist (3, 14). This enzyme, as well as its cytosolic equivalent, has been termed α-amylase in analogy with the cereal grain endo-amylase. However, the properties of the leaf enzymes differ from those of the classical amylases (15), and thus the designation endo-amylase should be preferred. The debranching enzyme of spinach leaves, some properties of which are detailed in Table II, showed identical behavior irrespective of whether a protein extract of whole leaves or of isolated chloroplasts was used as starting material for the purification (13). Thus, if an extra-chloroplastic R-enzyme exists, its activity is indistinguishable from that of the chloroplastic enzyme. Partially inactivated R-enzyme was reactivated by thiols. Depending on the proportion of inactive enzyme in the preparation, the purified debranching enzyme could thus be stimulated up to a 5-fold degree by the addition of DTT (13). No such effect was observed with the chloroplastic endo- and exo-amylase, respectively.

The physiological role of the extrachloroplastic starch-degrading leaf enzymes remains an enigma. When the pH-optima of the amylolytic system (ca. 6) and of phosphorylase are taken into consideration, it is to be expected that the cytosolic amylases will always be more active than those of the chloroplast, since the pH of the cytosol appears to be rather more acidic than that of the stroma (5). Therefore, significant amounts of oligo- or polyglucans cannot be expected to exist outside the chloroplast. The enigma then originates from the finding that all enzyme activities involved in starch synthesis, namely, ADP-glucose pyrophosphorylase, starch synthase, and starch-branching enzyme, have been found exclusively localized with the chloroplast (14), and thus the ability to produce a homoglucan outside that organelle lacks any substantiation.

The Maltose Transport System of Spinach Chloroplasts. Herold et al. (7), have shown that maltose is able to penetrate the chloroplast envelope at low rates (1.3 µmol h^{-1} mg^{-1} Chl at an external concentration of 2 mM). When investigating maltose uptake by isolated spinach chloroplasts in some detail, we were able to confirm this observation, but found rates 3-fold higher than those quoted by these authors. Since maltose is readily incorporated into maltodextrins, prolonged uptake results in an apparent accumulation within the chloroplast (Fig. 5). However, a true

Table II. **Purification and Properties of Spinach R-enzyme**

	Purification (from 13)		
Source of Protein	Chloroplasts	Leaf Extract − DTT	Leaf Extract + 10 mM DTT
Specific activity of purified enzyme (U/mg protein)	15.9	41.5	97.1
Recovery (%)	34.0	6.8	14.0
Purification (-fold)	408	3500	7000

Properties	
<u>Substrate Specificity (%)</u>	
Pullulan	100
Soluble starch	37
Amylopectin	17.5
Amylose	−
Various glycogens	−
pH-Optimum	5 − 5.5
Kinetic data	
"K_m" pullulan mg/ml	0.78
"K_m" amylopectin mg/ml	7.0
"V_{max}" pullulan U/mg Protein	69
"V_{max}" amylopectin U/mg Protein	22
Molecular properties	
Monomeric	Yes
MW (kD)	102 − 110
Competitive inhibitor	cyclohepta-amylose
"K_i" mg/ml	0.0009
Activation/reactivation	thiols, dithiols

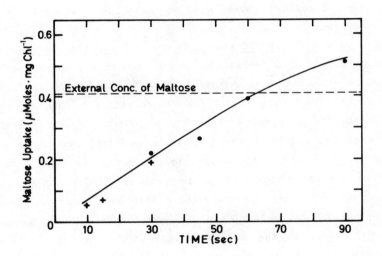

FIG. 5. Time kinetics of [U-^{14}C]maltose uptake into the sorbitol impermeable space (stroma) of intact spinach chloroplasts. The isolated chloroplasts were greater than 90% intact. The procedure described by Beck et al. (1), for uptake studies with ascorbate was employed with the following modifications: (i) [U-^{14}C]maltose (10 mM) was used instead of labeled ascorbate, (ii) DTT was replaced by nonlabeled ascorbate (2 mM), and (iii) the sorbitol content of the packed chloroplasts was determined by HPLC using a Polypore H column (10 μm, 4.6 x 30 mm; Brownlee Labs, Santa Clara, CA) and 0.013 M H_2SO_4 (at 50°C) as solvent.

accumulation indicative of active transport could not be confirmed. As has been found with the glucose translocator (21), maltose uptake at relatively low external concentrations exhibited saturation kinetics with an apparent K_m of 25 mM, and a V_{max} of about 40 µmol h^{-1} mg^{-1} Chl (Fig. 6), but at external concentrations higher than 10 mM, the uptake increased markedly and showed no saturation. Assuming a chloroplastic starch content of 50 to 90 µatoms mg^{-1} Chl (see above) and amylolysis as the only mode of starch breakdown, the stromal maltose concentration would ultimately amount to 100 to 200 mM. Although it is clear that this would never be the case, the calculation indicates that stromal concentrations in a millimolar range are not unlikely to occur, and that considerable physiological importance may thus be ascribed to the maltose transporting system of the chloroplast envelope with respect to carbohydrate export into the cytoplasm.

Maltose versus Triose-phosphate Export: Energetic Considerations. The export of carbon from transitory starch into the cytoplasm serves two major purposes--the supply of energy for the metabolism of the cell, and the provision of the carbon source for the formation of sucrose which is then transported to the entire plant by the phloem. Phosphorolysis preserves part of the energy of the glycosidic bond and thus saves one ATP per hexose as compared to amylolysis; hexose monophosphate can directly enter the oxidative pentose phosphate pathway (30). However, since hexose monophosphate cannot permeate the chloroplast envelope, one ATP is required to produce two molecules of triose-phosphate (via fructose-1,6-bisphosphate) which, by the action of the phosphate translocator, can then enter the cytoplasm. For the metabolism of triose-phosphate by cytoplasmic glycolysis, no further input of energy is required. However, sucrose synthesis from triose-phosphate requires the investment of one UTP in order to produce UDP-glucose. In this case, three energy equivalents would thus be necessary for the formation of sucrose, namely, two provided by the chloroplast for export and one from the cytoplasm for the establishment of the glycosidic bond.

On the other hand, three energy equivalents are also required if maltose is excreted from the chloroplast and hydrolyzed by cytoplasmic maltase to free glucose, which is then utilized to form sucrose commencing with the hexokinase reaction. In the case of this pathway, the necessary

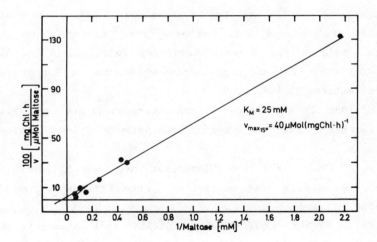

FIG. 6. Concentration dependence of maltose uptake into the sorbitol impermeable space of intact spinach chloroplasts. The period of uptake was 20 sec. Other conditions as described for Figure 5.

energy must be provided exclusively by the cytoplasm. If maltose phosphorylase rather than maltase is involved in the cleavage of maltose, synthesis of sucrose from maltose would then require only two energy equivalents.

These considerations lead to the conclusion that, from an energetic viewpoint, amylolytic starch degradation is not less favorable than phosphorolysis with regard to the conversion of starch to sucrose.

Acknowledgment—The author is grateful to Dr. G. Acker and Dr. P. Pongratz, for providing the electron micrographs, and to Dr. R. Scheibe and Dr. P. Ziegler for critical reading of the manuscript (all University of Bayreuth). The work was supported by the Deutsche Forschungsgemeinschaft.

LITERATURE CITED

1. BECK E, A BURKERT, M HOFMANN 1983 Uptake of L-ascorbate by intact spinach chloroplasts. Plant Physiol 73:41-45
2. BECK E, P PONGRATZ, I REUTER 1981 The amylolytic system of isolated chloroplasts and its role in the breakdown of assimilatory starch. In G Akoyunoglou, ed, Photosynthesis Proc. 5th Int. Congr. Photosynth., Balaban Int Science Services, Philadelphia pp 529-538
3. CHANG CW 1982 Enzymic degradation of starch in cotton leaves. Phytochemistry 21:1263-1269
4. GRUPE M 1982 Gelelectrophoretische Untersuchung des amylolytischen Systems bei Spinat- und Erbsenchloroplasten. Diploma Thesis Bayreuth
5. HEBER U, HW HELDT 1981 The chloroplast envelope: Structure, function, and role in leaf metabolism. Annu Rev Plant Physiol 32: 139-168
6. HELDT HW, K WERDAN, M MILOVANCEV, G GELLER 1971 Alkalinization of the chloroplast stroma caused by light-dependent proton flux into the thylakoid space. Biochim Biophys Acta 314:224-241
7. HEROLD A, RC LEEGOOD, PH McNEIL, SP ROBINSON 1981 Accumulation of maltose during photosynthesis in protoplasts isolated from spinach leaves treated with mannose. Plant Physiol 67:85-88

8. KAKEFUDA G, HS DUKE 1984 Electrophoretic transfer as a technique for the detection and identification of plant amylolytic enzymes in polyacrylamide gels. Plant Physiol 75:278-280
9. KRISMAN CR 1962 A method for the colorimetric estimation of glycogen with iodine. Anal Biochem 4:17-23
10. LEVI C, M GIBBS 1976 Starch degradation in isolated spinach chloroplasts. Plant Physiol 57:933-935
11. LEVI C, J PREISS 1978 Amylopectin degradation in pea chloroplast extracts. Plant Physiol 61:218-220
12. LINDEN JC, W TANNER, O KANDLER 1974 Properties of glucosyltransferase and glucan transferase from spinach. Plant Physiol 54:752-757
13. LUDWIG I, P ZIEGLER, E BECK 1984 Purification and properties of spinach leaf debranching enzyme. Plant Physiol 74:856-861
14. OKITA TW, E GREENBERG, DN KUHN, J PREISS 1979 Subcellular localization of the starch degradative and biosynthetic enzymes of spinach leaves. Plant Physiol 64:187-192
15. OKITA TW, J PREISS 1980 Starch degradation in spinach leaves. Isolation and characterization of the amylase and R-enzyme of spinach leaves. Plant Physiol 66:870-876
16. PEAVEY DG, M STEUP, M GIBBS 1977 Characterization of starch breakdown in isolated spinach chloroplasts. Plant Physiol 60:305-308
17. PONGRATZ P, E BECK 1978 Diurnal oscillation of amylolytic activity in spinach chloroplasts. Plant Physiol 62:687-689
18. PREISS J 1982 Regulation of the biosynthesis and degradation of starch. Annu Rev Plant Physiol 33:431-454
19. PREISS J, C LEVI 1979 Metabolism of starch in leaves. In M Gibbs, E. Latzko, eds, Photosynthesis II, Encyclopedia of Plant Physiology (NS) Vol 6, Springer, Berlin Heidelberg, pp 282-312
20. SARKO A, H-CH WU 1978 The crystal structures of A-, B-, and C-polymorphs of amylose and starch. Starch/Stärke 30:73-78
21. SCHÄFER G, U HEBER, HW HELDT 1977 Glucose transport into spinach chloroplasts. Plant Physiol 60:286-289
22. SCHILLING N, O KANDLER 1975 α-glucose-1-phosphate, a precursor in the biosynthesis of maltose in higher plants. Biochem Soc Trans 3:985-987

23. STEUP M 1981 Purification of chloroplast α 1,4-glucan phosphorylase from spinach leaves by chromatography on Sepharose-bound starch. Biochim Biophys Acta 659:123-131
24. STEUP M, H ROBENEK, M MELKONIAN 1983 In vitro degradation of starch granules isolated from spinach chloroplasts. Planta 158:428-436
25. STEUP M, C SCHÄCHTELE 1981 Mode of glucan degradation by purified phosphorylase forms from spinach leaves. Planta 153:351-361
26. STEUP M, C SCHÄCHTELE, E LATZKO 1980 Separation and partial characterization of chloroplast and nonchloroplast α-glucan phosphorylases from spinach leaves. Z Pflanzenphysiol 96:365-374
27. STEUP M, C SCHÄCHTELE, E LATZKO 1980 Purification of a nonchloroplastic α-glucan phosphorylase from spinach leaves. Planta 148:168-173
28. STITT M, T ap REES 1980 Carbohydrate breakdown by chloroplasts of Pisum sativum. Biochim Biophys Acta 627:131-143
29. STITT M, PV BULPIN, T ap REES 1978 Pathway of starch breakdown in photosynthetic tissue of Pisum sativum. Biochim Biophys Acta 544:200-214
30. STITT M, HW HELDT 1981 Physiological rates of starch breakdown in isolated intact spinach chloroplasts. Plant Physiol 68:755-761

Roles of Pyrophosphate and Fructose-2,6-Bisphosphate in Regulating Plant Sugar Metabolism

CLANTON C. BLACK, NANCY W. CARNAL, and NACHMAN PAZ

The reciprocating flow of carbon between hexoses and trioses is a central process in all plant cells. Regulating the flow in either direction is poorly understood. Though we have known for decades that plants utilize sugars in almost countless ways; e.g. by synthesizing hexoses, carbon is directed into dissaccharides, trisaccharides, or six carbon alcohols for translocation throughout a plant. In the plastid, trioses may be used to synthesize starch or, alternatively, trioses may flow out of the plastid to the cell cytoplasm for metabolic utilization as in amidation, oxidation, or hexose synthesis processes. Our understanding of carbon flow between hexoses and trioses currently is undergoing a vigorous re-examination because of two recent findings. First was the realization that pyrophosphate (PPi) is an energy source and phosphate donor for glycolytic carbon flow in plants (4) and, second, the finding that a new hexose, fructose-2,6-bisphosphate (Fru-2,6-P_2) is present and active (15) as a modulator of sugar metabolism (7, 10, 11, 21, 22). Here we will concentrate upon data from our work which led us to an interpretative model for the roles of PPi and Fru-2,6-P_2 in plant sugar metabolism. More comprehensive reviews on PPi and Fru-2,6-P_2 have been published recently (2, 17).

PLANT PHOSPHOFRUCTOKINASES

We know today that plants contain two sets of enzyme systems to catalyze the reversible interconversion of Fru-6-P and Fru-1,6-P_2. From Fru-1,6-P_2 carbon can flow either into trioses or into other hexoses. The phosphorylation of Fru-6-P is catalyzed either by a PPi-dependent or an ATP-dependent enzyme in plants as shown in Figure 1. We wish to concentrate on the PPi-dependent phosphofructophosphotransferase (PFPase) because it has several unique characteristics relative to the ATP-phosphofructokinase (PFKase). Some unique characteristics of plant PFPase are: (i) the use of PPi as an energy source and as its phosphate-containing substrate or product; (ii) the reaction is readily reversible at physiological substrate concentrations (Fig. 1); (iii) an intracellular localization of PFPase in the cytoplasm in contrast to PFKase which is both in the plastid and in the cytoplasm; and (iv) a sensitivity to the presence of $Fru-2,6-P_2$, whereas the plant PFKase is insensitive to $Fru-2,6-P_2$. The plant PFKase catalyzes a physiological irreversible synthesis of $Fru-1,6-P_2$; in fact, a second enzyme, $Fru-1,6-P_2$, 1-phosphatase (FbPase), is necessary to hydrolyze the 1-phosphate to form Fru-6-P. However, FbPase also is sensitive to $Fru-2,6-P_2$, indeed inhibited by $Fru-2,6-P_2$ (7, 10, 21).

Characterization of the Pineapple PPi-Phosphofructophosphotransferase. When the PFPase was found in plants, its proposed role was in glycolysis, as well as in obtaining energy from PPi to drive sugar metabolism (4). Pineapple leaves, which conduct Crassulacean acid metabolism, were the source of the PFPase studied initially. During her thesis research, N. Carnal (1, 3-5) purified and characterized this leaf enzyme. From a kinetic study, it was learned that the substrate responses are parallel as shown in Figure 2. Other work demonstrated that the product/substrate inhibition patterns are competitive or mixed with respect to each other. Collectively, these responses are consistent with a substituted, ping-pong, enzyme reaction mechanism in the direction toward $Fru-1,6-P_2$ formation as drawn in Figure 3 (3).

Early work on the kinetics of the pineapple PFPase in the direction of $Fru-1,6-P_2$ formation noted an inhibition at high, millimolar, levels of PPi (4). The reasons for a substrate inhibition were not known then, but later we learned that the enzyme dissociates in the presence of high PPi levels (23, 24). About 1982, we came to realize that the newly detected activator of the plant PFPase (15), $Fru-2,6-P_2$, would relieve the high PPi inhibition as shown in Figure 4 (3).

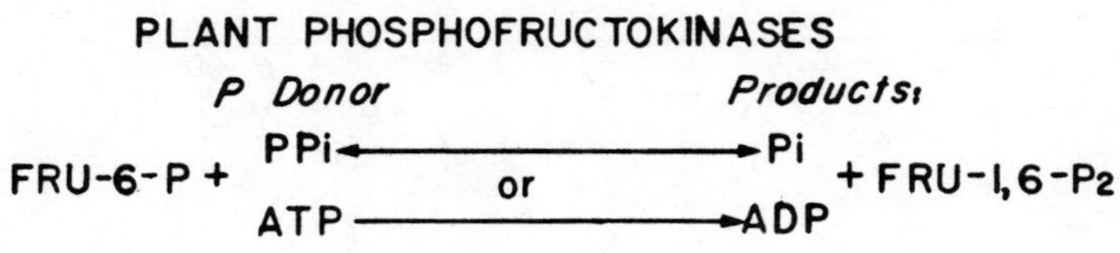

PLANT PHOSPHOFRUCTOKINASES

$$\text{FRU-6-P} + \begin{array}{c} \text{PPi} \\ \text{ATP} \end{array} \xrightarrow[\text{or}]{} \begin{array}{c} \text{Pi} \\ \text{ADP} \end{array} + \text{FRU-1,6-P}_2$$

(P Donor → Products)

FIG 1. Reactions catalyzed by the two types of plant phosphofructokinases, namely, a PPi-dependent phosphofructophosphotransferase, and an ATP-dependent phosphofructokinase.

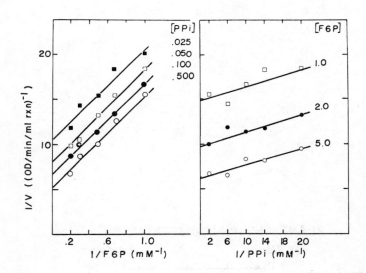

FIG 2. Double reciprocal plots of reaction velocity <u>versus</u> PPi concentration (right frame) or Fru-6-P concentration (left frame) at the indicated concentrations of PPi and Fru-6-P, respectively. The patterns of parallel lines are consistent with the ping-pong mechanism in Figure 3.

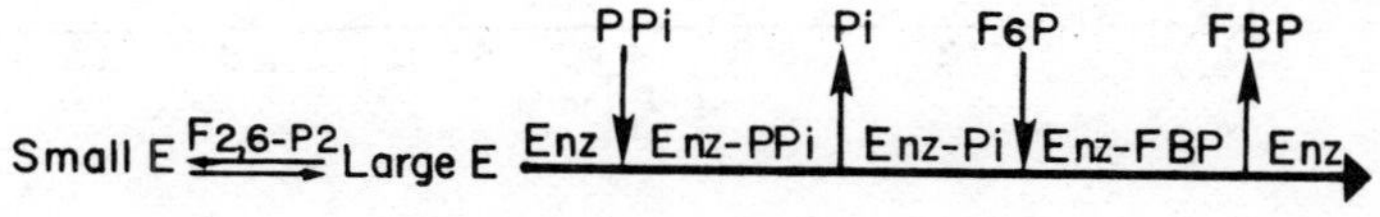

```
                                  PPi      Pi    F6P      FBP
                                   ↓       ↑      ↓        ↑
Small E ⇌^{F2,6-P2} Large E   Enz | Enz-PPi | Enz-Pi | Enz-FBP | Enz →
```

FIG. 3. Overall mechanism (ping-pong) for the molecular conversion of the small form of PPi-phosphofructotransferase into the large form of the enzyme by Fru-2,6-P_2 in the glycolytic direction.

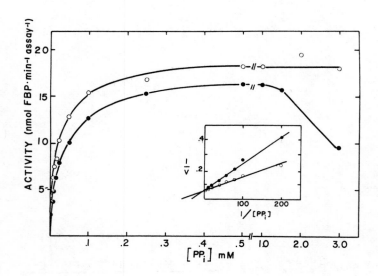

FIG. 4. Reaction velocity of PPi-phosphofructophosphotransferase as a function of PPi concentration. Experiments were run in the absence (●) or presence (o) of 0.5 µM Fru-2,6-P$_2$. Note the relief of PPi inhibition by Fru-2,6-P$_2$.

The pineapple PFPase is very active in the absence of exogenously added Fru-2,6-P$_2$ (4), but the enzyme is sensitive to Fru-2,6-P$_2$. In a study of the pineapple, PFPase kinetics in both the forward and reverse directions (as written in Fig. 1), in the presence and absence of Fru-2,6-P$_2$, the following kinetic parameters were obtained. With Fru-2,6-P$_2$ present at 0.5 µM, the K_m^{app} (Fru-6-P) dropped 6-fold from 2.2 mM (minus Fru-2,6-P$_2$) to 0.35 mM; simultaneously the K_m^{app} (PPi) dropped from 30 µM to 16 µM; and the K_m^{app} (Mg^{+2}) dropped from 120 µM to 6 µM. But the V_{max} was changed only slightly, (perhaps 10 to 15% increases seen occasionally) by Fru-2,6-P$_2$ additions. In the reverse direction, the K_m^{app} (Fru-1,6-P$_2$) dropped from 21 to 11 µM in the presence of Fru-2,6-P$_2$, but neither the K_m^{app} (Pi) of 0.73 mM, nor the V_{max} were influenced by Fru-2,6-P$_2$. In the forward direction, the K_a (Fru-2,6-P$_2$) is near 4 nM. The V_{max} forward is two times the V_{max} reverse (3). Collectively, these data lead to the conclusion that the leaf PFPase functioned both in the glycolytic and the gluconeogenic flow of carbon between hexoses and trioses; but how Fru-2,6-P$_2$ functioned remained to be elucidated along with further understandings of PPi metabolism.

REGULATION OF PLANT ENZYMES BY FRUCTOSE-2,6-BISPHOSPHATE

Our clue to how Fru-2,6-P$_2$ functioned in regulating PFPase came from work on the enzyme in developing pea seedlings where M.-X. Wu noted that for a 3- to 5-d period the pea enzyme was insensitive to added Fru-2,6-P$_2$; but earlier or later in development the enzyme was sensitive (23). By gel filtration chromatography (Sephadex G-200), two forms of the enzyme were isolated, one sensitive and the other insensitive to Fru-2,6-P$_2$. Furthermore, we found the two forms were interconvertible by binding with Fru-2,6-P$_2$ and other ligands such as PPi. Thus we came to realize that the enzyme existed in two interconvertible molecular forms whose activity depended upon association and dissociation characteristics regulated by Fru-2,6-P$_2$ (23, 24). The two forms of PFPase have sedimentation coefficients of 12.7 S (large form) and 6.3 S (small form). The enzyme apparently has two subunits, α and β, and the large form is α$_2$β$_2$ and the small is αβ (25). The large form dissociates in the absence of Fru-2,6-P$_2$, or in the presence of ligands such as high levels of PPi (24); but

Fru-2,6-P$_2$ will reverse this dissociation and restore activity upon association (Fig. 4).

Thus, the kinetic properties of the two forms are different, as illustrated in Figure 5 with the pea enzyme. In the glycolytic direction, the large form of the enzyme is only slightly stimulated by Fru-2,6-P$_2$ but, in marked contrast, the activity of the small form is greatly increased by Fru-2,6-P$_2$ with a K_a of 15 nM. In the gluconeogenic direction, neither form of PFPase is sensitive to Fru-2,6-P$_2$, even at levels up to 100 μM (24). With the pea seedling enzyme at optimum substrate assay conditions, the ratio of the glycolytic to the gluconeogenic activity for the large and small forms are 3.3 and 0.27 respectively. Therefore, the large form is more active in glycolysis while the small form is more active in gluconeogenesis.

Several other plant enzymes are reported to be sensitive to Fru-2,6-P$_2$ (2, 14, 17), but these will not be discussed here except to state that FbPase is inhibited by Fru-2,6-P$_2$ at Ki values ranging from 0.1 to 100 μM when assayed with various plant sources of the enzyme (7, 10, 21). Clearly, the ability to inhibit this gluconeogenic enzyme also is a regulatory site in triose to hexose conversions which potentially involved Fru-2,6-P$_2$.

PYROPHOSPHATE CONTENT OF PLANT TISSUES AND ITS METABOLISM

A holistic view of the major role for PPi in biochemistry is that PPi is synthesized during polymer (e.g. starch, glycogen, sucrose, cellulose, protein, DNA, RNA) synthesis and hydrolyzed to form inorganic phosphate; thereby providing a thermodynamic pull, favoring polymer synthesis. Hence near zero, or a very low, level of PPi should be present in living tissues. Until recently, no PPi had been measured in plant tissue; however, in our laboratory, D. Smyth realized the PFPase could be used to measure PPi. Using the PPi-dependent PFPase, two laboratories simultaneously reported the presence of PPi at levels ranging from 9 to 39 nmol/g fresh weight with pea and corn tissues (8, 16). The exact cellular localization of PPi in plant cells is unknown, but assuming it is in the cytoplasm (10% of a cell volume), these levels calculate to 100 to 200 μM PPi, well above the K_m (PPi) of PFPase. We conclude that PPi

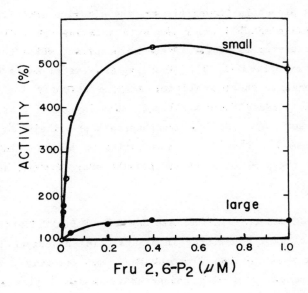

FIG. 5. Activity of the large or small form of PPi-phosphofructokinase in the glycolytic direction as a function of Fru-2,6-P_2 concentration.

is present in plants as an appreciable pool and its presence provides the phosphate-donating substrate and the energy source for PFPase. In addition, the presence of such a substantial energy source implies that plant energetics must be re-evaluated (8, 16). We have initiated these ideas by demonstrating that plant glycolysis, in converting Fru-6-P to pyruvate, is driven by PPi, without ATP, and that added Fru-2,6-P_2 stimulates the PPi-dependent glycolysis (19). In another area of research, for several years it has been contended that mitochondria and chloroplasts synthesize PPi; this work has been extended recently (13).

In the classical view of PPi metabolism, it is widely thought that inorganic pyrophosphatases readily hydrolyze PPi to inorganic phosphate. Plant inorganic pyrophosphatases are poorly characterized; even so, they would directly compete with PFPase for PPi. Even with the limited information on plant pyrophosphatases, it appears that two major forms are present, one with an alkaline pH optimum and the other with an acid optimum. With pineapple leaves (1, 3), the alkaline activity (pH 8.5) is near 22.8 µmol/min/mg Chl, while the acid activity (pH 5.8) is near 1.7. The alkaline activity is mostly in the chloroplast, while the acid activity is in the cytoplasm (1, 3). Assuming the cytoplasm is alkaline, the acid pyrophosphatase there would not compete strongly for PPi with the PFPase. We have indeed found a PPi pool near 14 nmol^{-1} g fresh weight in pineapple leaves (16). Hence, pyrophosphatases likely do not compete well with PFPase for the PPi in plant tissues because they either are sequestered in the plastid or their activities are minimal at cytoplasmic pH values.

FRUCTOSE-2,6-BISPHOSPHATE CONTENT OF PLANT TISSUES

The detection of Fru-2,6-P_2 in plants, plus an enzyme for the addition and hydrolysis of the 2-phosphate, came directly from repeating similar experiments originally conducted with animal tissues (6, 9, 11, 15, 18). Figure 6 outlines these reactions for Fru-2,6-P_2 metabolism and, in addition, lists the common plant metabolites which partially modulate the 2-kinase and the 2-phosphatase. These subjects are under intensive study currently, because many questions remain about Fru-2,6-P_2 metabolism (2, 17).

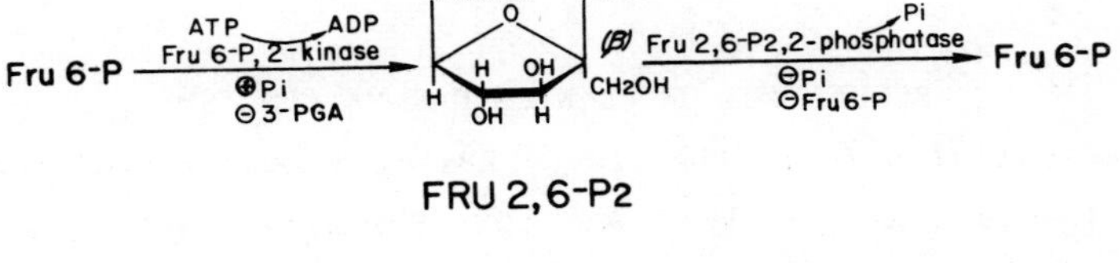

Fru 6-P $\xrightarrow[\substack{\oplus Pi \\ \ominus 3\text{-PGA}}]{\substack{ATP \quad ADP \\ \text{Fru 6-P, 2-kinase}}}$ [Fru 2,6-P2 structure: CH2OPO3, OPO3, H, OH, CH2OH, OH, H] $\xrightarrow[\substack{\ominus Pi \\ \ominus Fru\,6\text{-}P}]{\text{Fru 2,6-P2,2-phosphatase} \quad Pi}$ Fru 6-P

FRU 2,6-P2

FIG. 6. The structure of fructose-2,6-bisphosphate and the modulators of the phosphorylation and dephosphorylation enzymes which synthesize and hydrolyze Fru-2,6-P_2 in plants. ⊕ = stimulation; ⊖ = inhibition.

Here, we only wish to consider the daily changes in Fru-2,6-P_2 levels in plants and then how these changes imply an involvement of Fru-2,6-P_2 in the regulation of hexose and triose conversions. The sensitivity of the plant PFPase to Fru-2,6-P_2 is again used as the assay (22). N. Paz, in studying the daily patterns of total Fru-2,6-P_2 which can be extracted from pea seedling tissues, found the changes shown in Figure 7. In leaf tissues there is a drop in Fru-2,6-P_2 level as illumination begins and then a rise through the day with little detectable change occurring in the night. Quite similar day kinetics have been reported for spinach leaves (20). But in root tissue, the leaf pattern is reversed in that Fru-2,6-P_2 levels increase through the day followed by a sharp decrease early in the night (Fig. 7). In other work with pea-stem tissue, which one can think of as the translocation connection between the source (leaves) and the sink (roots), the day pattern is similar to the leaf and the night pattern is similar to the root data shown in Figure 7. Therefore, each type of plant tissue exhibits a unique pattern for regulating its content of Fru-2,6-P_2. It is clear that plant tissues can change the total amount of Fru-2,6-P_2 several fold (12, 20, 22, Fig. 7); though the concentrations at the cellular sites of enzyme localization are unknown currently.

However, we postulate these results imply that sugar metabolism is regulated by Fru-2,6-P_2 each day as a normal course of events in each plant tissue. Accepting this, even though we do not know what triggers the synthesis or degradation of Fru-2,6-P_2 in plants (2, 17), we can obtain a reasonably unified concept of how Fru-2,6-P_2 regulates sugar metabolism.

REGULATION OF PLANT SUGAR METABOLISM

In the initial work on the plant PFPase, it was postulated that the enzyme was involved in glycolysis and that PPi served as an energy source in plant metabolism (4). In plants it now has been demonstrated: (i) that a PPi-dependent PFPase is quite active and is found in essentially all types of tissues; (ii) that PPi is present at quantities sufficient to serve as a substrate; (iii) that Fru-2,6-P_2 regulates the molecular interconversion of PFPase into a large and small form; (iv) that PPi serves as

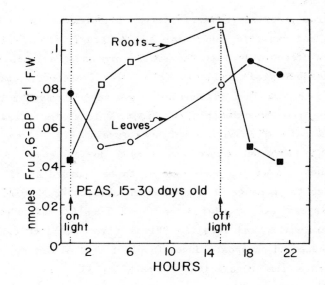

FIG. 7. Diurnal changes in the levels of Fru-2,6-P_2 in leaf and root tissue from pea seedlings. Closed symbols are for data with dark tissue and open symbols are for tissues in the light.

an energy source for the glycolytic conversion of Fru-6-P to pyruvate; (v) that Fru-2,6-P_2 is present in various tissues; and (vi) the levels of Fru-2,6-P_2 in plant tissues vary each day (Fig. 7). We wish to combine these results with other observations to present a model (Fig. 8) for the regulation of hexose metabolism by Fru-2,6-P_2 at the level of Fru-6-P interconversion with Fru-1,6-P_2 catalyzed by PFPase and Fru-1,6-P_2, 1-phosphatase (2, 17).

In brief, PPi is considered to be the phosphate donor for the phosphorylation of Fru-6-P. If the levels of Fru-2,6-P_2 rise in a tissue (as with the day stage of root tissue in Fig. 7), then the PFPase associates and is activated toward glycolysis. Simultaneously in the reverse direction, the activity catalyzed mostly by the small form of PFPase is not active and the Fru-1,6-P_2, 1-phosphatase also is inhibited. Thus, raising the Fru-2,6-P_2 level in a plant tissue favors glycolysis; e.g. in the day root tissue of Figure 7, which likely is importing sucrose, glycolysis is favored each day.

In contrast, during the early day period, leaves as in Figure 7 are synthesizing and exporting sucrose. As the Fru-2,6-P_2 level drops (Fig. 7) the PFPase dissociates into the small form and the inhibition of Fru-1,6-P_2, 1-phosphatase is relieved (Fig. 8). Therefore, gluconeogenesis is favored in the day leaf tissues which are actively synthesizing sucrose using trioses from the chloroplast. Figure 8 is a simplified model linking plant PPi metabolism with Fru-2,6-P_2 levels (Fig. 7) to regulate the flow of carbon between hexose and triose during sugar metabolism in various tissues. Even so, much remains to be learned about sugar metabolism; such as the message or signal which elicits Fru-2,6-P_2 metabolism; how other enzymes are regulated by Fru-2,6-P_2 (2, 17), how do plants synthesize, maintain, and localize PPi in vivo; how important is PPi as an energy source and phosphate donor in plant metabolism; or, how do plant cells coordinate the regulation of both an ATP- and a PPi-dependent phosphofructose kinase (Fig. 1) in their cytoplasm.

ACKNOWLEDGMENTS--This work was graciously supported by the National Science Foundation through grant DMB 84-06331 to C.C.B. Dr. H. M. Vines generously assisted in preparing the figures.

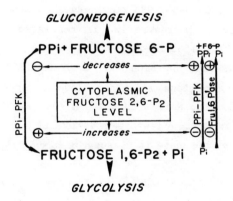

FIG. 8. Overall model for the use of pyrophosphate as an energy source and for the regulation of plant sugar metabolism. By changing cytoplasmic Fru-2,6-P_2 concentrations, the level of hexose interconversions are altered by PPi-PFPase and Fru-1,6-P_2, 1-phosphatase. ⊕ = stimulation; ⊖ = inhibition.

LITERATURE CITED

1. BLACK CC, NW CARNAL, WK KENYON 1982 Compartmentation and the regulation of CAM. In IP Ting, M Gibbs, eds, Crassulacean Acid Metabolism. Am Soc of Plant Physiol, Rockville, MD, pp 51-68
2. BLACK CC, DA SMYTH, M-X WU 1985 Pyrophosphate-dependent glycolysis and regulation by fructose 2,6-bisphosphate in plants. In PW Ludden, JE Burris, eds, Nitrogen Fixation and CO_2 Metabolism. Elsevier Sci Pub Co, pp 361-370
3. CARNAL NW 1984 Pyrophosphate: 6-phosphofructokinase in plants: Discovery, characterization, and an examination of its role in carbohydrate metabolism. PhD thesis. University of Georgia, Athens, GA 305 pp
4. CARNAL NW, CC BLACK 1979 Pyrophosphate-dependent 6-phosphofructokinase, a new glycolytic enzyme in pineapple leaves. Biochem Biophys Res Commun 86:20-26
5. CARNAL NW, CC BLACK 1983 Phosphofructokinase activities in photosynthetic organisms. The occurrence of pyrophosphate-dependent 6-phosphofructokinase in plants and algae. Plant Physiol 71:150-155
6. CSÉKE C, BB BUCHANAN 1983 An enzyme synthesizing fru-2,6-bisphosphate occurs in leaves and is regulated by metabolite effectors. FEBS Lett 155:139-142
7. CSÉKE C, NF WEEDEN, BB BUCHANAN, K UYEDA 1982 A special fructose bisphosphate functions as a cytoplasmic regulatory metabolite in green leaves. Proc Natl Acad Sci USA 79:4322-4326
8. EDWARDS J, T ap REES, PM WILSON, S MORRELL 1984 Measurement of the inorganic pyrophosphate in tissues of Pisum sativum L. Planta 162:188-191
9. FURUYA E, M YOKOYAMA, K UYEDA 1982 An enzyme that catalyzes hydrolysis of fructose-2,6-bisphosphate. Biochem Biophys Res Commun 105:264-170
10. GOTTSCHALK ME, T CHATTERJEE, I EDELSTEIN, F MARCUS 1982 Studies on the mechanism of interaction of fructose-2,6-bisphosphate with fructose-1,6-bisphosphatase. J Biol Chem 237:8016-8020
11. HERS H-G, L HUE, E VAN SCHAFTINGEN 1982 Fructose-2,6-bisphosphate. Trends Biochem Sci 7:329-331

12. HUBER SC, DM BICKETT 1984 Evidence for control of carbon partitioning by fructose-2,6-bisphosphate in spinach leaves. Plant Physiol 74:445-447
13. KOWALCZYK S, P MASLOWSKI 1984 Oxidation-linked formation of inorganic pyrophosphate in maize shoot mitochondria. Biochim Biophys Acta 766:570-575
14. KRUGER NJ, E KOMBRINK, H BEEVERS 1983 Pyrophosphate fructose 6-phosphate phosphotransferase in germinating castor bean seedlings. FEBS Lett 153:409-412
15. SABULARSE DC, RL ANDERSON 1981 D-fructose-2,6-bisphosphate: A naturally occurring activator for inorganic pyrophosphate: D-fructose-6-phosphate 1-phosphotransferase. Biochem Biophys Res Commun 103:848-855
16. SMYTH DA, CC BLACK 1984 Measurement of the pyrophosphate content of plant tissues. Plant Physiol 75:862-864
17. SMYTH DA, CC BLACK 1984 The discovery of a new pathway of glycolysis in plants. What's New Plant Physiol 15:13-16
18. SMITH DA, M-X WU, CC BLACK 1984 Phosphofructokinase and fructose 2,6-bisphosphatase activities in developing corn seedlings (Zea mays L.). Plant Sci Lett 33:61-70
19. SMITH DA, M-X WU, CC BLACK 1984 Pyrophosphate and fructose 2,6-bisphosphate effects on glycolysis in pea seed extracts. Plant Physiol 76:316-320
20. STITT M, R GERHARDT, B KÜRZEL, HW HELDT 1983 A role for fructose 2,6-bisphosphate in the regulation of sucrose synthesis in spinach leaves. Plant Physiol 72:1139-1141
21. STITT M, G MEISKES, H-D SÖLING, HW HELDT 1982 On a possible role of fructose 2,6-bisphosphate in regulating photosynthetic metabolism in leaves. FEBS Lett 145:217-222
22. VAN SCHAFTINGEN E, B LEDERER, R BARTRONS, H-G HERS 1982 A kinetic study of pyrophosphate:fructose-6-phosphate phosphotransferase from potato tubers. Application to a microassay of fructose 2,6-bisphosphate. Eur J Biochem 129:191-195

23. WU M-X, DA SMYTH, CC BLACK JR 1983 Fructose 2,6-bisphosphate and the regulation of pyrophosphate-dependent phosphofructokinase activity in germinating pea seeds. Plant Physiol 73:188-191
24. WU M-X, DA SMYTH, CC BLACK JR 1984 Regulation of pea seed pyrophosphate-dependent phosphofructokinase. Evidence for the interconversion of two molecular forms as a glycolytic regulatory mechanism. Proc Natl Acad Sci USA 81:5051-5055
25. YAN T-F J, M TAO 1984 Multiple forms of pyrophosphate:D-fructose-6-phosphate 1-phosphotransferase from wheat seedlings. J Biol Chem 259:5087-5092

Characterization of Fructose-1,6-Bisphosphatase Activity from *Synechococcus leopoliensis*

ERWIN LATZKO, KLAUS-PETER GERBLING, and MARTIN STEUP

In photoautotrophic organisms, the fructose-1,6-bisphosphate-1,phosphatase (FbPase) reaction has multiple functions, associated with the reductive and oxidative pentose phosphate cycle and with gluconeogenesis. A variety of regulatory mechanisms have been postulated to exert selective control of the enzyme activity in these reaction sequences. In higher plant cells, the regulation of FbPase is facilitated by the relatively high degree of compartmentation and by the existence of compartment-specific FbPases. The plastids and the cytoplasm contain FbPase forms which substantially differ in their kinetic and regulatory properties (1, 4, 10, 11). For the chloroplast FbPase form, several interconvertible kinetic states have been proposed (5-7). The in vitro interconversions depend upon the presence of reductant/oxidant, substrate, divalent cations or upon alkalization/acidification of the medium. It is thought that these effects may have important implications for the in vivo control of chloroplast FbPase and, thus, of the reductive pentose phosphate cycle. Unlike the chloroplastic FbPase, the cytoplasmic enzyme form exhibits a high sensitivity toward inhibitors such as AMP and fructose-2,6-bisphosphate (Fru-2,6-P_2) (9, 11).

Although procaryotic cells exhibit a low degree of compartmentation, multiple FbPase forms have recently been described for Synechococcus leopoliensis (2). In crude extracts of Synechococcus cells, FbPase activity was separated into two components by isoelectric focusing, followed by enzyme activity staining (Fig. 1, a and b). The forms, designated as form A and form B, exhibited isoelectric points in the ranges of 4.7 to

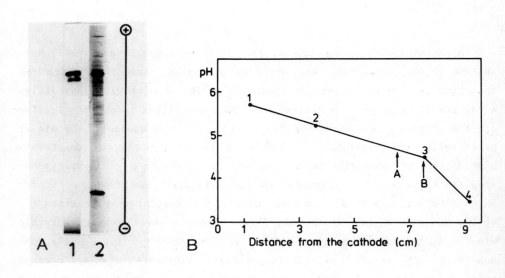

FIG. 1. a, Fructose-1,6-bisphosphate-1,phosphatase zymogram of a crude extract of Synechococcus leopoliensis after isoelectric focusing. Isoelectric focusing was from pH 3.5 to 10.0. FbPase activity was detected by a coupled enzyme assay in which the NADPH formed then reduced tetrazolium bromide, causing the formation of insoluble formazan. Lane 1: FbPase activity staining; lane 2: protein staining. b, Isoelectric point determination of FbPase, forms A and B (pH 4.0-6.5). Isoelectric point markers were: 1, bovine carbonic anhydrase (I.E.P. 5.85); 2, β-lactoglobulin A (5.2); 3, soybean trypsin inhibitor (4.55); 4, amyloglucosidase (3.5). A, B: FbPase forms.

4.8 and of 4.5 to 4.6, respectively. In addition to isoelectric focusing, the two FbPase forms were also resolved by nondenaturing polyacrylamide gel electrophoresis and by preparative chromatofocusing (2). No in vitro interconversions were detected between these forms.

Form B, which probably represents the major proportion of the total FbPase activity, has been purified by the procedure summarized in Table I. The purification procedure included an ammonium sulfate precipitation, hydrophobic chromatography, chromatofocusing, affinity chromatography, and gel filtration. Forms A and B were separated by chromatofocusing which, however, resulted in a low recovery of form A. In those experiments in which form A was compared with form B, the form A preparation obtained after chromatofocusing was used. The purification of form B resulted in a recovery of 57% of the total extractable FbPase activity.

The form B preparation was homogeneous, as revealed by denaturing polyacrylamide gel electrophoresis. The monomer exhibited a molecular mass of 38 to 40 kD (see Fig. 5a, lane III, and Fig. 5b). The molecular mass of form B, in its native state, was determined by gel filtration and by gradient electrophoresis. The values obtained were strongly dependent upon the presence or absence of Fru-1,6-P_2 and Mg^{2+}. When gel filtration of form B was performed in the presence of either Fru-1,6-P_2 or Mg^{2+}, the enzyme exhibited an apparent molecular mass of 53 to 59 kD, indicating a dimeric enzyme subform (Fig. 2a). The same apparent molecular mass was obtained when both Fru-1,6-P_2 and Mg^{2+} were omitted (Fig. 2b). However, if gel filtration was performed in the presence of both Fru-1,6-P_2 and Mg^{2+}, a tetrameric subform with an apparent molecular mass of 122 to 129 kD was obtained (Fig. 2b). Comparable experiments performed with form A of FbPase did not result in a shift of the apparent molecular mass (Fig. 2c).

The Fru-1,6-P_2/Mg^{2+}-dependent dimer-tetramer conversion of form B could also be demonstrated with crude extracts. However, the conversion appeared to be incomplete (Fig. 2d). This is probably due to the presence of form A, which remains in the low molecular mass subform (cf. Fig. 2c).

The dimer-tetramer conversion of form B was observed over a wide range of experimental conditions (enzyme concentrations 85-500 µg/ml; buffer: Hepes or Tris, and pH values ranging between 6.0 and 8.0; data

Table I. Purification of Fructose-1,6-Bisphosphate-1, Phosphatase Form B from Synechococcus leopoliensis

Fraction	Volume (ml)	Activity (mU)	Protein (mg)	Specific Activity (mU/mg protein)	Purification (-fold)	Recovery (%)
1. Crude extract	74	587	148	3.95	-	100
2. Ammonium sulfate (form A and B)	81	747	113	5.6	1.4	127
3. Hydrophobic chromatography on hexyl-Agarose (form A and B)	83	617	13	47.4	12	105
4. Chromatofocusing on poly-buffer exchange gel						
(form A)	33	1.5	1	1.5	0.4	2.6
(form B)	28	344	1.9	181	46	59
5. Affinity chromatography on Fru-1,6-P_2-Sepharose (form B)	36	307	0.61	500	127	52
6. Gel filtraton on Sephadex G 100 (form B)	39	332	0.59	562	142	57

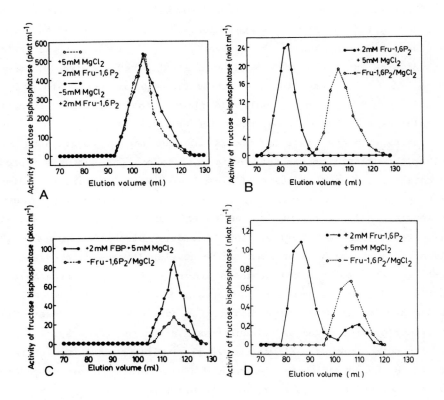

FIG. 2. Gel filtration of fructose-1,6-bisphosphate-1,phosphatase, forms A and B. Gel filtration (see Ref. 2) was at low pressure in 20 mM Hepes-NaOH containing 5 mM DTT (Kat = 10^6 x U = mol/min). a (top, left panel), Purified form B; gel filtration performed at pH 8.0 in the presence of either Fru-1,6-P_2 or Mg^{2+}. b (top, right panel), Purified form B; gel filtration performed in the absence of Fru-1,6-P_2/Mg^{2+} at pH 8.0 (o------o) or in the presence of Fru-1,6-P_2/Mg^{2+} at pH 7.0 (●------●). c (bottom, left panel), FbPase form A, gel filtration performed at pH 8.0 in the absence or in the presence of Fru-1,6-P_2/Mg^{2+}. d (bottom, right panel), Crude extract; conditions as in (b).

not shown). Further, the reversibility of the dimer-tetramer conversion could be demonstrated most conveniently by high-pressure gel filtration, a procedure which reduces the separation time (Fig. 3).

The results obtained by both types of gel filtration were confirmed by nondenaturing polyacrylamide gel electrophoresis performed both in the presence and in the absence of Fru-1,6-P_2/Mg^{2+} (Fig. 4). In the presence of Fru-1,6-P_2/Mg^{2+}, the form B of FbPase exhibited a molecular mass of 130 to 140 kD, representing the tetrameric subform (Fig. 4b). In the absence of Fru-1,6-P_2/Mg^{2+}, only the dimeric subform, with an apparent molecular mass of 65 to 70 kD, was detected (Fig. 4a).

The experiments described so far do not permit conclusions as to whether the dimeric or the tetrameric subform or both subforms represent the catalytically active state of form B of FbPase. In order to distinguish among these possibilities, the different subforms were fixed by covalent cross-linking using bis(sulfosuccinimidyl)-suberate (3, 8). This imidoester reacts with primary or secondary aliphatic amines. Thus, oligomeric subforms of a protein can be covalently cross-linked and fixed together. The cross-linked products are resistant to SDS-treatment (8); therefore, denaturing gel electrophoresis can be used to determine the subform species present under different conditions. The monomer, dimer, and trimer of BSA served as molecular mass standards (Fig. 5).

The data shown in Figure 5 clearly indicate that the tetrameric subform is the active state of form B of FbPase. This subform is generated from a dimer which has to be reduced before the Fru-1,6-P_2/Mg^{2+}-dependent aggregation to a tetramer occurs.

These conclusions were confirmed by kinetic measurements with the various states of form B of FbPase (Fig. 6). When form B of FbPase was added to the assay mixture as a tetramer, maximal activity was reached after a very short lag of 10 to 20 s. When the dithiothreitol (DTT)-treated dimer was added to the assay, and the reaction was initiated by Fru-1,6-P_2/Mg^{2+}, the lag extended to 4 min. If, however, the dimer was added without a prior reduction by DTT, the time required to reach maximum activity was increased to 8 to 10 min.

In summary, we have demonstrated the presence of two forms of FbPase, forms A and B, in the blue-green alga, Synechococcus leopoliensis. Form

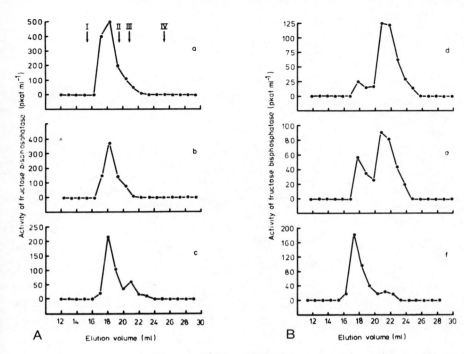

FIG. 3. Interconversion of purified fructose-1,6-bisphosphate-1, phosphatase form B. Experiments were by gel filtraton (HPLC) on TSK G 3000 column. Equilibration and elution were with 150 mM Tris-HCl pH 7.0 and 5 mM DTT. γ-Globulin (I: 150,000), BSA (II; 67,000), ovalbumin (III; 45,000) and myoglobin (IV: 178,000) were used as molecular mass standards (Kat = 10^6 x U). a, Form B, preincubated in 150 mM Tris-HCl pH 7.0, 5 mM DTT, 2 mM Fru-1,6-P_2, 5 mM Mg^{2+}, was passed through the column equilibrated with the same buffer. b, Form B was freed of Fru-1,6-P_2/Mg^{2+} by a passage through a Sephadex G-25 column equilibrated with 150 mM Tris-HCl pH 7.0, 5 mM DTT and was immediately applied to the TSK G 3000 column, previously equilibrated with the same buffer. c, Form B was incubated for 15 min at 4°C in 150 mM Tris-HCl pH 7.0, 5 mM DTT prior to gel filtration (HPLC). All other conditions as in (b). d, 120-min incubation at 4°C; other conditions as in (b). e, As in (d), but 2 mM Fru-1,6-P_2, 5 mM Mg^{2+} were added to form B after the 120-min incubation prior to gel filtration (HPLC). f, As in (e), but form B was incubated for 30 min at 4°C after the addition of 2 mM Fru-1,6-P_2, 5 mM Mg^{2+} prior to gel filtration (HPLC).

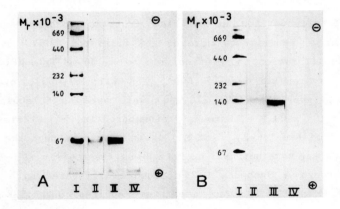

FIG. 4. Nondenaturing polyacrylamide gel electrophoresis of purified fructose-1,6-bisphosphate-1,phosphatase form B. FbPase (5 µg protein per lane) was run in the absence (a, left panel) or in the presence (b, right panel) of Fru-1,6-P_2/Mg^{2+}. I, molecular mass standards; II-IV, FbPase form B; II, protein staining; III, activity staining (complete mixture); IV, activity staining mixture, but Fru-1,6-P_2 omitted.

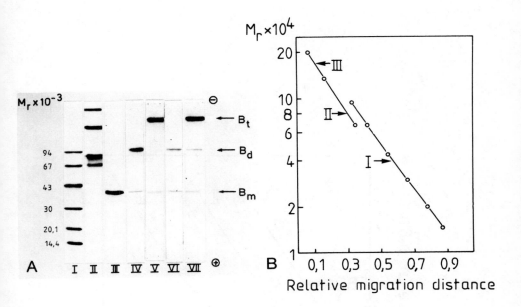

FIG. 5. Denaturing polyacrylamide gel electrophoresis of fructose-1,6-bisphosphate-1,phosphatase form B after cross-linking. The cross-linking was with bis(sulfosuccinimidyl)suberate (BS). a (left panel), Lane I, noncross-linked molecular mass standards; II, cross-linked molecular mass standards (monomer, dimer, and trimer of BSA); III, FbPase without BS treatment; IV, FbPase incubated with DTT without Fru-1,6-P_2/Mg^{2+} prior to treatment with BS; V, FbPase incubated with DTT followed by incubation with Fru-1,6-P_2/Mg^{2+} prior to BS treatment; VI, FbPase incubated with cystamine followed by incubation with Fru-1,6-P_2/Mg^{2+}; VII, FbPase incubated in the following order: Fru-1,6-P_2/Mg^{2+}, cystamine, and finally BS treatment. b (right panel), Determination of the apparent molecular mass of FbPase form B after cross-linking with BS. Standards as in 5a. I, II, III, monomeric, dimeric, and tetrameric subforms of FbPase form B.

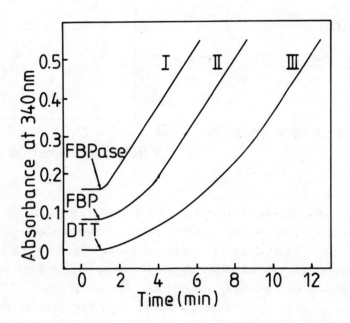

FIG. 6. Reaction rate of fructose-1,6-bisphosphate-1,phosphatase form B. The activity was measured after different preincubations (30 min at 4°C). I, preincubation with DTT, plus Fru-1,6-P_2/Mg^{2+}; II, preincubation with DTT without Fru-1,6-P_2/Mg^{2+}; III, preincubation without both DTT and Fru-1,6-P_2/Mg^{2+}. Aliquots of the preincubation mixtures were added to the assays; reaction was initiated as indicated.

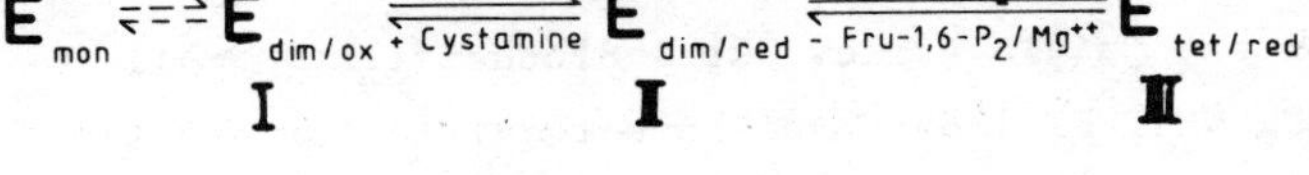

$$E^{ci}_{mon} \rightleftharpoons E^{ci}_{dim/ox} \xrightarrow[\text{+ Cystamine}]{\text{+ DTT}} E^{ci}_{dim/red} \xrightleftharpoons[\text{- Fru-1,6-}P_2/Mg^{++}]{\text{+ Fru-1,6-}P_2/Mg^{++}} E^{ca}_{tet/red}$$
$$\phantom{E^{ci}_{mon} \rightleftharpoons E^{ci}_{dim/ox}} \text{I} \phantom{\xrightarrow[\text{+ Cystamine}]{\text{+ DTT}}} \text{I} \phantom{\xrightleftharpoons[\text{- Fru}]{\text{+ Fru}}} \text{III}$$

FIG. 7. Scheme of the interconversions of various subforms of form B of fructose-1,6-bisphosphate-1,phosphatase. E^{ci} and E^{ca}, catalytically inactive and active enzyme states; E_{mon}, E_{dim}, and E_{tet}, the monomeric, dimeric, and tetrameric subforms of FbPase after cystamine (E_{ox}) or DTT (E_{red}) preincubation.

B was purified to homogeneity. This form exists in at least three different states (monomer, dimer, tetramer); only one of which is catalytically active. The interconversion of these states is outlined in the scheme in Figure 7.

It is not known whether a pool of the monomeric enzyme exists under native conditions or whether there is a reversible transition between the monomer and state I. However, the data clearly show that states I to III are readily interconvertible. Although the kinetics of the interconversion could not be followed in detail for technical reasons, the results of Figures 3 and 5 indicated that the transitions from state I to III are completed in less than 30 min, even at 4°C. At 30°C, the entire sequence from I to III took place in ca. 10 min (Figure 6).

LITERATURE CITED

1. BUCHANAN BB 1980 Role of light in the regulation of chloroplast enzymes. Annu Rev Plant Physiol 31:341-374
2. GERBLING K-P, M STEUP, E LATZKO 1984 Electrophoretic and chromatographic separation of two fructose-1,6-bisphosphatase forms from Synechococcus leopoliensis. Arch Microbiol 137:109-114
3. GIEDROC DP, D PUETT, N LING, JV STAROS 1983 Demonstration by covalent cross-linking of a specific interaction between β-endorphin and calmodulin. J Biol Chem 258:16-19
4. KELLY GJ, E LATZKO 1984 Photosynthesis. Carbon metabolism: On land and at sea. Prog in Botany 46:68-93
5. PRADEL J, JM SOULIÉ, J BUC, JC MEUNIER 1981 On the activation of fructose-1,6-bisphosphatase of spinach chloroplasts and the regulation of the Calvin cycle. Eur J Biochem 113:507-511
6. ROSA L, FR WHATLEY 1984 Conditions required for the rapid activation in vitro of the chloroplast fructose-1,6-bisphosphatase. Plant Physiol 75:131-137
7. SOULIÉ JM, J BUC, JC MEUNIER, J PRADEL, J RICARD 1981 Molecular properties of chloroplastic thioredoxin f and the photoregulation of the activity of fructose-1,6-bisphosphatase. Eur J Biochem 119:417-502

8. STAROS JV 1982 N-hydroxysulfosuccinimide active esters: Bis(N-hydroxysulfosuccinimide) esters of two dicarboxylic acids are hydrophylic, membrane-impermeant, protein cross-linkers. Biochemistry 21:3950-3955
9. STITT M, G MIESKES, H-D SÖLING, HW HELDT 1982 On a possible role of fructose 2,6-bisphosphate in regulating photosynthetic metabolism in leaves. FEBS Lett 145:217-222
10. ZIMMERMANN G, GJ KELLY, E LATZKO 1976 Efficient purification and molecular properties of spinach chloroplast fructose 1,6-bisphosphatase. Eur J Biochem 70:361-367
11. ZIMMERMANN G, GJ KELLY, E LATZKO 1978 Purification and properties of spinach leaf cytoplasmic fructose-1,6-bisphosphatase. J Biol Chem 253:5952-5956

Pyrophosphate and the Glycolysis of Sucrose in Higher Plants

TOM AP REES, SUSAN MORRELL, JEFFREY EDWARDS, PATRICIA M. WILSON, and JOHN H. GREEN

Sucrose is the prime respiratory substrate in higher plants, and glycolysis is the dominant pathway of carbohydrate oxidation. In this article, we consider whether the entry of carbon from sucrose into plant glycolysis involves the use of inorganic pyrophosphate as a source of energy. One of the axioms of biochemistry is that the driving force for the synthesis of macromolecules is provided by the splitting of nucleoside triphosphates to give PPi that is rapidly hydrolyzed by inorganic pyrophosphatase (18). It has been suggested that pyrophosphatase breaks down PPi almost as soon as it is formed (29). Some years ago, Wood, O'Brien and Michaels (31) pointed out that the driving force for biosynthesis would still be provided if PPi were removed by means other than hydrolysis, namely by using it as a source of energy. They also reviewed evidence that in some organisms, PPi may serve as an energy source. However, these organisms are highly specialized, and it was not clear whether the above use of PPi was widespread. Evidence that it is, has been provided by the discovery of the general occurrence of PPi: fructose 6-phosphate phosphotransferase (PFPase) in plants (9, 19). This discovery has led to suggestions that plants may use PPi to synthesize $Fru-1,6-P_2$ for glycolysis (21, 25).

The essential difficulty with the above hypothesis is that the reaction catalyzed by PFPase is readily reversible <u>in vitro</u>, $\Delta G = -2.93$ kJmol^{-1}. This suggests that <u>in vivo</u>, the enzyme is as capable of catalyzing the conversion of $Fru-1,6-P_2$ to $Fru-6-P$ as it is of acting glycolytically. Up to date, the vast majority of plant tissues reported to contain PFPase are either photosynthetic (9) or gluconeogenic (19). In such tissues, the flux from Fru-1,6-P_2 to Fru-6-P would exceed glycolysis, and it is not obvious that the enzyme functions glycolytically rather than in the reverse direction. The fact that PFPase can replace ATP-dependent phosphofructokinase (PFKase) in cell-free glycolysis (26) is consistent with a glycolytic role for the former, but does not prove it. As PFPase is reversible, it is just as likely that it could replace fructose bisphosphate-1,phosphatase (FbPase) in cell-free gluconeogenesis.

Studies of partially purified preparations of PFPase, recently reviewed (3), have revealed important features of the enzyme, but not its role in vivo. The enzyme is specific for PPi. Estimates of K_m^{app} for PPi are low, 10 to 30 μM, as are those for Fru-1,6-P_2, 20 to 90 μM. Higher values have been obtained for Fru-6-P, 0.3 to 2.2 mM, and Pi, 0.7 to 0.8 mM. PPi excepted for the moment, none of these values are below the concentrations likely to exist in vivo. Perhaps the most noteworthy property of PFPase is its stimulation by Fru-2,6-P_2. Studies with the enzyme from castor bean endosperm (17) showed that in the glycolytic direction activation by Fru-2,6-P_2 is hyperbolic, enhanced by Fru-6-P and diminished by Pi. K_a for Fru-2,6-P_2 is low, but is markedly affected by Pi. At 5 mM Fru-6-P, the addition of 5 mM Pi increased K_a for Fru-2,6-P_2 from 12 to 123 nM. In the other direction, K_a is higher and is increased by increasing Pi. However, the complexity of the interactions between PFPase, its substrates, and Fru-2,6-P_2 is such that the differential response of the forward and back reactions to Fru-2,6-P_2 is not a definitive pointer to the direction that the reaction proceeds in vivo.

EXPERIMENTAL APPROACH

We adopted the following approach in an attempt to discover whether PPi served as an energy source by acting as a substrate for PFPase in plant glycolysis. First, we determined whether there was enough PPi present and whether the enzyme was generally distributed in nonphotosynthetic and nongluconeogenic tissues. Then we compared the maximum catalytic activity for PFPase with that of PFKase, and with the rate of respiration, during the differentiation of roots of 5 to 7-d-old seedlings of Pisum sativum L. cv. Kelvedon Wonder, and during the development of the club of the spadix of Arum maculatum L. This was done to discover if there was sufficient PFPase to mediate glycolysis, and if there was any relationship between PFPase and the rate of glycolysis.

In the above context, comparisons between measurements of enzyme activities are only meaningful if the measurements have been shown to reflect the maximum catalytic activities of the tissues analyzed. Unless we say otherwise, the measurements of enzymes were made with assays in which the pH, and the concentration of the components, had been optimized, and where activity had been shown to be linear with respect to time and amount of extract assayed. Losses of activity during extraction were investigated with recovery experiments, where pure samples of the enzyme were available, and with recombination experiments where they were not (invertase and sucrose synthase). In recovery experiments, we prepared duplicate samples, extracted one in the usual way, and the other in the same way except that a measured amount of the pure enzyme was included in the extraction medium. The difference between the activities in the two samples is expressed as a percentage of the activity of added enzyme to give an estimate of the recovery of the latter. The values we obtained were within 15% of the amounts added, and most were within 10%. In recombination experiments, we prepared three samples from two different tissues, one of one, one of the other, and one a mixture of known weight of the two. The activity in the mixture was always within 11% of the value predicted from the measurements made on the separate components. These results suggest that our estimates of maximum catalytic activity were not seriously affected by differential inactivation of enzymes during extraction and assay.

The following is evidence that interconversion of two forms of PFPase (32), or of PFPase and PFKase (7), did not occur to any significant extent during preparation of the extracts (also see Dennis et al, this symposium, p. 134). First, recovery of PFPase and PFKase was good. Second, the activities of both enzymes in freshly prepared extracts kept at 2°C remained constant for up to 1 h. Third, dilution of the endogenous metabolites during homogenization would have resulted in the extracts containing too low a concentration of compounds, like $Fru-2,6-P_2$, said to promote the above interconversions. Fourth, the activities of both enzymes varied independently, not inversely. Finally, when a sample of pea roots was freeze-clamped and put straight into extraction buffer, the activity of PFPase was the same as that found in a duplicate sample extracted in the conventional way.

For measurements of substrates, tissues were freeze-clamped, and then killed as described by ap Rees _et al_. (4). Losses during these procedures were investigated by preparing duplicate samples; one was treated as above, the other similarly except that measured amounts of the compound to be assayed were added to the tissue immediately after it had been freeze-clamped. The difference between the amounts assayed in the two extracts was used to estimate the recovery of the added compound. These estimates were within 17% of the amounts added. Details of the procedures and assays used are given in the accompanying references.

PPi Content of Plant Tissues. The hypothesis that PFPase acts glycolytically requires the presence of appreciable amounts of PPi in the cytosol. The dependence of PFPase on PPi offered a means of measuring PPi in unfractionated plant extracts. If PPi is present, it should give rise to Fru-1,6-P_2 which could then be measured by coupling to NADH oxidation via aldolase, triose-phosphate isomerase, and glycerol-3-phosphate dehydrogenase (13). When solutions of PPi were assayed, we found that the relationship between PPi taken up and NADH oxidized was linear at least up to 25 nmol PPi/ml of reaction mixture. Amounts of PPi as low as 1 nmol could be assayed. In applying the method to plant tissues, the major difficulty is to inactivate any pyrophosphatase in the extract by a means that does not itself cause chemical hydrolysis of PPi. The recoveries shown in Table I indicate that 1.41 M $HClO_4$ was satisfactory for the tissues that we examined: it might not be in tissues with higher pyrophosphatase activities. Evidence that the NADH oxidation that we measured was due to PPi is provided by the fact that addition of pyrophosphatase to the neutralized $HClO_4$ extracts abolished the oxidation.

The data in Table I provide evidence that three quite different tissues contain appreciable amounts of PPi. The absolute amounts are comparable to those of known glycolytic intermediates (see Table IV) and to estimates of PPi for pea and maize tissues reported by Smyth and Black (25). The concentration of PPi _in vivo_ is more difficult to assess. If

Table I. Pyrophosphate Content of Plant Tissues

Embryos of 300 to 400 mg fresh weight, the apical 2 cm of roots of 5 to 7-d-old seedlings, and clubs from prethermogenic spadices were taken as described previously (4, 13). Values are means ± SE from the number of samples shown in parentheses.

Tissue	PPi Content	Recovery of Added PPi
	nmol/g fresh weight	%
Developing embryos of *Pisum sativum*	9.4 ± 1.2 (11)	90 ± 3 (6)
Root apices of *Pisum sativum*	8.9 ± 0.5 (6)	102 ± 3 (5)
Club spadix of *Arum maculatum*	27.5 ± 3.4 (6)	99

we make the assumption that the PPi is in the cytosol, and that the latter is 10% of the volume of the tissue, then the concentration would be 100 to 300 μM, well in excess of K_m^{app} of PFPase for PPi.

PPi-phosphofructose Phosphotransferase, ATP-phosphofructokinase and Respiration. If PFPase is a gluconeogenic enzyme, we should not expect to find it in any quantity in tissues in which there is no apparent net flux from Fru-1,6-P_2 to Fru-6-P. Such a flux occurs in photosynthetic and gluconeogenic tissues, and may occur in nonphotosynthetic tissues that are converting sucrose to starch (20). Accordingly, we measured the activity of PFPase in a number of nonphotosynthetic, nongluconeogenic tissues that contain very little or no starch (Table II). The values for sugar beet and onion are minimal as the assays were not optimized for these two tissues. Appreciable activities of PFPase, capable of sustaining significant rates of respiration, were found in all tissues. These observations are consistent with a glycolytic role for the enzyme, and suggest that its principal function is not catalysis of a net flux from Fru-1,6-P_2 to Fru-6-P.

Table II. Activities of PPi-dependent Phosphofructophosphotransferase in Nonphotosynthetic, Nongluconeogenic, Nonstarchy Tissues

Activity was measured as Fru-1,6-P_2 formation as described in Ref. 4. Values represented as in Table I.

Tissue	PFPase Enzyme Activity
	μmol/g fresh weight min
Roots of Lolium temulentum	1.36 ± 0.06 (7)
Storage tissue from root of red beet	0.15 ± 0.01 (5)
White base of leaf of Allium porrum	0.88 ± 0.14 (7)
Storage tissue from root of sugar beet	0.18, 0.20
Storage tissue of Allium cepa	0.10

Next, we investigated whether the maximum catalytic activity of PFPase correlated with respiration during the differentiation of pea roots (Table III). We also present measurements for PFKase, as one of the main difficulties in accepting a glycolytic role for PFPase is the impressive body of evidence that plant glycolysis depends upon PFKase (28). Tissues at different stages of differentiation were obtained by dissecting the root into apex, cortex, and stele as described previously (30). All regions of the root examined contain enough of either enzyme to catalyze the observed rate of respiration. The maximum catalytic activities of both enzymes vary during differentiation, but do so independently. If it is remembered that the activity of glycolysis relative to that of the oxidative pentose phosphate pathway decreases at 5 mm from the root apex (14), then it can be argued that there is a general correlation between PFKase and the rate of respiration during the differentiation of the root. The activity of PFPase does not correlate with the rate of respiration. In general, PFPase activity is highest in the tissues where biosynthesis is likely to be most marked. These measurements are consistent with a glycolytic role for PFPase, but the lack of any clear

Table III. Activities of PPi-phosphosfructophosphotransferase, ATP-phosphofructokinase, and Respiration in Different Parts of the Root of 5 to 6-d-old Seedlings of Pisum sativum

PFPase was measured as described in reference 4, other data from references 24 and 30. Values for enzymes are means ± SE from at least 5 samples.

Region of Root		PFPase (PPi-dependent)	PFKase (ATP-dependent)	CO_2 Production
Apical 6 mm		297 ± 9	95 ± 3	54
6-26 mm from apex:	stele	259 ± 33	46 ± 2	30
	cortex	136 ± 9	52 ± 2	36
26-46 mm from apex:	stele	200 ± 25	50 ± 2	40
	cortex	66 ± 10	46 ± 3	36

correlation between PFPase and the rate of respiration affords no positive support for the hypothesis. The data do suggest that any contribution of PFPase to glycolysis is most marked in tissues characterized by extensive biosynthesis.

We tested the above hypothesis by carrying out experiments, similar to those done with pea roots, on the developing club of the spadix of Arum maculatum. The club passes through a series of arbitrarily defined developmental stages that we have described (1), and called α, β, γ, prethermogenesis and thermogenesis. The first four stages are characterized by a marked and continuous net synthesis of starch and protein that extends over a period of about 12 d. This phase of extensive biosynthesis is succeeded by the stage called thermogenesis in which the rate of respiration rises suddenly by as much as 60-fold, and the starch is almost completely oxidized to CO_2 via glycolysis and the Krebs cycle to support extensive thermogenesis. The distribution of ^{14}C from [^{14}C]glucose fed to club tissue at thermogenesis shows that little or no biosynthesis takes place at this stage. Thus, club development is characterized by a switch from a metabolism dominated by biosynthesis to one of almost complete catabolism.

Table IV presents estimates of PFPase, PFKase, and related metabolites during the development of the club of A. maculatum. We were unable to detect any significant change in the maximum catalytic activity of PFPase during the development of the club. We confirmed this conclusion by assaying the enzyme in the direction of Fru-6-P formation, as well as in the direction shown in Table IV. This behavior of PFPase contrasts markedly with that of PFKase and the other two glycolytic enzymes assayed. All these show substantial increases in maximum catalytic activity during development. As in pea root, there is no correlation between the rate of respiration and the activity of PFPase. From α stage to prethermogenesis there is sufficient PFPase to mediate glycolysis in vivo. At thermogenesis there is not. At thermogenesis, the rate of starch breakdown is so rapid that it may be used as a direct measure of the rate of glycolysis (6). We found that the average rate of glycolysis over a 30-min period during the development of the peak rate of respiration was 7.6 µmol hexose g^{-1} fresh weight min^{-1}. As the rate of respiration was rising throughout the period of measurement, the peak rate of glycolysis would have been in excess of the above value. We suggest that there is insufficient PFPase in the clubs for it to play a quantitatively important role in glycolysis at thermogenesis. This is not true of PFKase and the other enzymes of glycolysis assayed. The view that PFPase plays no major role in glycolysis at thermogenesis is strengthened by the behavior of PPi and Fru-2,6-P_2. Despite a massive and rapid rise in the rate of glycolysis, probably the greatest found in plants, neither metabolite showed any significant change between prethermogenesis and thermogenesis.

The data in Table IV allow estimation of the extent to which the reaction catalyzed by PFPase approaches equilibrium in vivo. For this purpose, we assume that PPi, Fru-6-P and Fru-1,6-P_2 are confined to the cytosol. We appreciate that a proportion of the last two compounds is likely to be in the amyloplast, but suggest that this proportion may well be minor as the stromal volume of these plastids appears to be very small (5). We have also assumed a cytosolic volume of 10% of that of the club. From the data in Table IV and the ΔG quoted earlier, we calculate that, on the above assumptions, the concentration of Pi that would have to be present in the cytosol if the PFPase reaction were at equilibrium would

Table IV. Enzyme Activities and Metabolite Contents during the Development of the Club of the Spadix of Arum maculatum

Data are adapted from references 1 and 4: PFPase was measured as $Fru-1,6-P_2$ formation. Values are means $\pm$ SE from at least 5 clubs.

	Stage of Development				
	α	β	γ	Prethermogenesis	Thermogenesis
Fresh weight of club (mg)	239 $\pm$ 8	616 $\pm$ 18	1169 $\pm$ 38	1346 $\pm$ 38	1205 $\pm$ 107
CO_2 production (μmol g^{-1} fresh weight min^{-1})	0.37 $\pm$ 0.01	0.48	—	0.61 $\pm$ 0.06	6–40
Enzyme activity (μmol g^{-1} fresh weight min^{-1})					
PFPase (PPi-dependent)	5.60 $\pm$ 0.21	5.24 $\pm$ 0.42	4.12 $\pm$ 0.30	4.02 $\pm$ 0.60	4.26 $\pm$ 0.58
PFKase (ATP-dependent)	1.15 $\pm$ 0.09	3.76 $\pm$ 0.73	12.28 $\pm$ 1.80	18.15 $\pm$ 1.40	30.01 $\pm$ 3.05
$Fru-1,6-P_2$ aldolase	3.74 $\pm$ 0.16	6.08 $\pm$ 0.97	8.57 $\pm$ 1.65	13.40 $\pm$ 0.72	14.58 $\pm$ 0.79
Glyceraldehyde 3-P dehydrogenase (NAD)	5.12 $\pm$ 0.89	11.98 $\pm$ 0.50	—	75.18 $\pm$ 10.61	—
Metabolite content (nmol g^{-1} fresh weight)					
$Fru-2,6-P_2$	0.56 $\pm$ 0.02	0.36 $\pm$ 0.03	0.30 $\pm$ 0.03	0.31 $\pm$ 0.05	0.46 $\pm$ 0.08
PPi	21.10 $\pm$ 1.8	36.9 $\pm$ 2.7	33.2 $\pm$ 3.9	27.5 $\pm$ 3.4	36.2 $\pm$ 2.0
Fru-6-P	35.8 $\pm$ 4.0	—	—	48.0 $\pm$ 4.6	70.0 $\pm$ 9.4
$Fru-1,6-P_2$	11.7 $\pm$ 2.2	—	—	33.8 $\pm$ 7.0	147.6 $\pm$ 22.5

be 1.4 and 0.74 mM, respectively, at α stage and thermogenesis. These values are a little lower than the current estimate of 6 mM for nonphotosynthetic cells of plants (22), but the difference is not marked. Thus, in <u>Arum</u> clubs, at least, it seems likely that PFPase catalyzes a near-equilibrium reaction <u>in vivo</u>.

Our results show that a wide range of plant tissues, that are neither photosynthetic nor gluconeogenic, contain sufficient PFPase to mediate the rates of glycolysis observed <u>in vivo</u>. A glycolytic role is also supported by the demonstration that the same types of plant cells contain appreciable amounts of PPi and Fru-2,6-P_2. However, a very striking feature of our results is the lack of any clear correlation between the maximum catalytic activity of PFPase and the rate of respiration. This is so marked in <u>Arum</u> clubs, that it seems clear that PFPase makes, at the most, a minor contribution to glycolysis at thermogenesis. The available data do not favor the view that PFPase makes a universal contribution to plant glycolysis, but are consistent with the enzyme contributing to glycolysis during biosynthesis. Thus, it is conceivable that there are two ways into plant glycolysis: via the PFPase with respect to biosynthesis and via PFKase where the major product of respiration is ATP. Although this hypothesis is consistent with the data that we have at present, the latter can be equally well explained by the hypothesis that PFPase acts to produce PPi required for the metabolism of sucrose.

<u>Sucrose Metabolism in Nonphotosynthetic Cells</u>. Reasons for regarding sucrose as the starting point of the metabolism of nonphotosynthesizing cells of plants have already been given (1, 2). The fates of this sucrose may be summarized as: storage, respiration, and conversion to either storage or structural polysaccharides. The division of sucrose among these fates must rest in the first instance on the relative activities of the three enzymes known to be generally distributed in plants and to be cabable of metabolizing sucrose. These are acid invertase, alkaline invertase, and sucrose synthase.

There is appreciable evidence that vacuolar acid invertase is responsible for hydrolysis of stored sucrose, and that this enzyme plays a key role in regulating whether sucrose is stored (2). The roles of the other two enzymes are less understood. Some years ago (1), it was permissible to suggest that sucrose synthase mediated direct conversion of sucrose to polysaccharide via UDP-glucose, and that alkaline invertase provided respiratory substrate in tissues that lacked acid invertase. Recent work strongly suggests that this simple view is wrong.

It now seems unlikely that a nucleotide sugar, formed directly from sucrose by sucrose synthase, is the immediate precursor of starch. First, ADP-glucose, not UDP-glucose appears to be the precursor of starch in nonphotosynthetic as well as photosynthetic cells (5). Second, sucrose synthase is confined to the cytosol, which almost certainly contains sufficient UDP to ensure that UDP-glucose, not ADP-glucose, is the product of sucrose synthase in vivo (20). These considerations prompt the hypothesis that sucrose synthase is involved in the conversion of sucrose to structural polysaccharides, the enzymes responsible for their synthesis having access to cytosolic UDP-glucose, and that starch synthesis and respiration depend upon the invertases. The work we now describe was done to test this hypothesis.

As experimental material we chose three developing tissues in which the conversion of sucrose to starch is a dominant feature of metabolism. These were potato tubers, the embryos of P. sativum (cv Greenshaft), and the club of A. maculatum. The following is evidence that in all three tissues synthesis of starch overshadowed that of structural polysaccharides. When [^{14}C]glucose was supplied to potato tubers, starch was heavily labeled and relatively little label was recovered in other polysaccharides (10). When the pea embryos were supplied with [^{14}C]glucose, 58% of the metabolized label was incorporated into insoluble compounds; 39% into starch, and 11% into protein (12). In the Arum clubs, over 90% of the label incorporated into insoluble material from [^{14}C]glucose was recovered in starch (Hargreaves, personal communication). If the above hypothesis is correct, we would expect all three tissues to contain sufficient invertase to supply starch synthesis and respiration, and relatively little sucrose synthase. Our estimates of the maximum catalytic activities of the enzymes show that the reverse was true (Table V). Comparable data have been reported for maize seeds (11). Measurement of starch accumulation during the development of the tubers referred to in Table V gave a value of 47 nmol hexose units g^{-1} fresh weight min^{-1}.

Table V. **Activities of Enzymes of Sucrose Metabolism in Potato Tubers, Pea Embryos, and Clubs of Arum maculatum**

All three organs were in mid-development when sampled. The tubers (cv Pentland Javelin), and embryos weighed 10 g, and 0.3 g, respectively; clubs were at β stage except those used for assay of sucrose synthase which were at γ stage. Assays were: invertases (23), PFPase (Table II), remaining enzymes (5).

	Activity		
Enzyme	Potato Tuber	Pea Embryo	Arum Club
	μmol g^{-1} fresh weight min^{-1}		
Acid invertase	0.002 $\pm$ 0.009	None detected	–
Alkaline invertase	0.007 $\pm$ 0.009	0.192 $\pm$ 0.008	–
Sucrose synthase	0.213 $\pm$ 0.033	2.683 $\pm$ 0.097	1.16 $\pm$ 0.21
UDP-glucose pyrophosphorylase	31.4 $\pm$ 1.3	57.33 $\pm$ 0.05	4.29 $\pm$ 0.94
PFPase	7.13 $\pm$ 0.74	1.547 $\pm$ 0.107	5.21 $\pm$ 0.42

This minimal estimate of the rate of synthesis of starch from sucrose is well within the capacity of sucrose synthase in the tuber, but far outside that of either invertase.

The above results make it very difficult to believe that sucrose synthase is only involved in the conversion of sucrose to structural polysaccharide. We suggest that in potato tubers and pea embryos, at least, this enzyme plays a dominant role in providing hexose phosphate for starch synthesis and respiration. The conversion of fructose, produced by sucrose synthase, to Fru-6-P could be readily accomplished by a cytosolic fructokinase such as that demonstrated by Tanner *et al.* (27). The difficulty lies in obtaining hexose phosphate from the UDP-glucose formed by sucrose synthase. This could be achieved by UDP-glucose pyrophosphorylase, which is located in the cytosol (20) and catalyzes a reversible reaction with an equilibrium relation (15) of:

$$\frac{[UTP][Glc-1-P]}{[PPi][UDP-glucose]} = 2.9 - 3.6$$

Two further considerations support this suggestion. The maximum catalytic activities of UDP-glucose pyrophosphorylase are extremely high (Table V). This might be expected if the enzyme were on the major route whereby incoming sucrose entered the metabolism of the cell. The high activity is very difficult to reconcile with the minor flux from hexose phosphates to structural polysaccharides via UDP-glucose. In addition, the amounts of UDP-glucose (5), PPi (13), Glc-1-P (8) and UTP (16) in developing embryos of pea, measured as in the accompanying references, were 324, 9.4, 70 and 175 nmol g^{-1} fresh weight. If it is assumed that these compounds are confined to the cytosol, then a mass action ratio of 4.2 is obtained for comparison with the above equilibrium constant. The similarity of the two suggests that the step is readily reversible in vivo. Thus, we envisage sucrose metabolism as:

Sucrose + UDP   ↔ Fructose + UDP-glucose

Fructose + ATP   → Fructose-6-phosphate

UDP-glucose + PPi   ↔ Glucose-1-phosphate + UTP

Glucose-1-phosphate   ↔ Glucose-6-phosphate

Our proposed route of sucrose metabolism requires a substantial supply of PPi. The main significance of PFPase in plant metabolism may be provision of this PPi. This hypothesis entails the use of Fru-1,6-P_2 by PFPase to produce PPi and Fru-6-P, with the latter re-entering glycolysis or the oxidative pentose phosphate pathway. The sensitivity of PFPase to Fru-2,6-P_2 might allow the latter to regulate the disposition of sucrose entering nonphotosynthetic cells. Thus the significance of Fru-2,6-P_2 in plants may lie in its regulation of sucrose metabolism, not glycolysis. Much of our knowledge of PFPase is consistent with this suggestion. The enzyme is as widely distributed as is the use of sucrose as the prime source of carbon and energy in nonphotosynthesizing cells. So is the observation that the maximum catalytic activity of PFPase generally exceeds that of PFKase and correlates with total demand for sucrose, rather than the rate of respiration. This would be expected if PFPase had to supply the complete needs of metabolism and PFKase only those of respiration. This situation would be reversed in tissues that had a high rate of respiration supported primarily by an endogenous supply of starch, as in the thermogenic clubs of Arum.

CONCLUSIONS

We suggest that there is sufficient PPi in plant tissues for it to act as an energy source. There is appreciable evidence that PPi does act in this way during the metabolism of the sucrose that is delivered to developing embryos of pea and tubers of potato, where PPi is envisaged as driving UDP-glucose pyrophosphorylase towards glucose 1-phosphate. We appreciate that there may be a range of ways of producing the PPi needed to metabolize sucrose, but we point out that one possible source is PFPase. The key to this last question is the direction of the net flux catalyzed by PFPase in vivo. The presently available evidence does not distinguish between the view that the enzyme is a source of PPi and the hypothesis that it provides an entry into glycolysis in tissues characterized by marked biosynthesis.

Acknowledgments—S. M. and J. E. thank the Potato Marketing Board and the Science and Engineering Research Council, respectively, for studentships.

LITERATURE CITED

1. AP REES T 1977 Conservation of carbohydrate by the nonphotosynthetic cells of higher plants. Symp Soc Exp Biol 31:7-32
2. AP REES T 1984 Sucrose metabolism. In DH Lewis, ed, Storage Carbohydrates in Vascular Plants. SEB Seminar Series, No. 19. Cambridge University Press, Cambridge, pp 53-73
3. AP REES T 1985 The organization of glycolysis and the oxidative pentose phosphate pathway in plants. In R Douce, DA Day, eds, Encyclopedia of Plant Physiology, New Series, Higher Plant Cell Respiration, Springer-Verlag, Berlin. In press
4. AP REES T, JH GREEN, PM WILSON 1985 Pyrophosphate: fructose 6-phosphate phosphotransferase and glycolysis in nonphotosynthetic tissues of higher plants. Biochem J. In press

5. AP REES T, M LEJA, FD MACDONALD, JH GREEN 1984 Nucleotide sugars and starch synthesis in spadix of <u>Arum maculatum</u> and suspension cultures of <u>Glycine max</u>. Phytochemistry 23:2463-2468
6. AP REES T, BW WRIGHT, WA FULLER 1977 Measurements of starch breakdown as estimates of glycolysis during thermogenesis by the spadix of <u>Arum maculatum</u> L. Planta 134:53-56
7. BALOGH A, JH WONG, C WÖTZEL, J SOLL, C CSÉKE, BB BUCHANAN 1984 Metabolite-mediated catalyst conversion of PFP and PFK: a mechanism of enzyme regulation in green plants. FEBS Lett 169:287-292
8. BERGMEYER HU, G MICHAL 1974 D-Glucose-1-phosphate. <u>In</u> HU Bergmeyer, ed, Methods of Enzymatic Analysis, Ed 2, Vol 3, Verlag Chemie, Weinheim, pp 1233-1242
9. CARNAL NW, CC BLACK 1983 Phosphofructokinase activities in photosynthetic organisms. The occurrence of pyrophosphate-dependent 6-phosphofructokinase in plants and algae. Plant Physiol 71:150-155
10. DIXON WL, T AP REES 1980 Carbohydrate metabolism during cold-induced sweetening of potato tubers. Phytochemistry 19:1653-1655
11. ECHEVERRIA E, T HUMPHRIES 1984 Involvement of sucrose synthase in sucrose catabolism. Phytochemistry 23:2173-2178
12. EDWARDS J 1985 Conversion of sucrose to starch in pea embryos. PhD thesis, University of Cambridge
13. EDWARDS J, T AP REES, PM WILSON, S MORRELL 1984 Measurement of the inorganic pyrophosphate in tissues of <u>Pisum sativum</u> L. Planta 162:188-191
14. FOWLER MW, T AP REES 1970 Carbohydrate oxidation during differentiation in roots of <u>Pisum sativum</u>. Biochim Biophys Acta 201:715-724
15. HANSEN RG, GJ ALBRECHT, ST BASS, LL SEIFERT 1966 UDP glucose pyrophosphorylase (crystalline) from liver. Methods Enzymol 8:248-253
16. KEPPLER D, G KARLFRIED, K DECKER 1974 Uridine-5'-triphosphate, uridine-5'-diphosphate and uridine-5'-monophosphate. <u>In</u> HU Bergmeyer, ed, Methods of Enzymatic Analysis, Ed 2, Vol 4, Verlag Chemie, Weinheim, pp 2172-2178

17. KOMBRINK E, NJ KRUGER, H. BEEVERS 1984 Kinetic properties of pyrophosphate: fructose-6-phosphate phosphotransferase from germinating castor bean endosperm. Plant Physiol 74:395-401
18. KORNBERG A 1962 On the metabolic significance of phosphorolytic and pyrophosphorolytic reactions. In M Kasha, B Pullman, eds, Horizons in Biochemistry. Academic Press, New York, pp 251-264
19. KRUGER NJ, E KOMBRINK, H BEEVERS 1983 Pyrophosphate: fructose 6-phosphate phosphotransferase in germinating castor bean seedlings. FEBS Lett 153:409-412
20. MACDONALD FD, T AP REES 1983 Enzymic properties of amyloplasts from suspension cultures of soybean. Biochim Biophys Acta 755:81-89
21. PREISS J 1984 Starch, sucrose biosynthesis and partition of carbon in plants are regulated by orthophosphate and triosephosphates. Trends Biochem Sci 9:24-27
22. REBEILLE F, R BLIGNY, J-P MARTIN, R DOUCE 1983 Relationship between the cytoplasm and the vacuole phosphate pool in Acer pseudoplatanus cells. Arch Biochem Biophys 225:143-148
23. RICARDO CPP, T AP REES 1970 Invertase activity during the development of carrot roots. Phytochemistry 9:239-247.
24. SMITH AM, T AP REES 1979 Effects of anaerobiosis on carbohydrate oxidation by roots of Pisum sativum. Phytochemistry 18:1453-1458
25. SMYTH DA, CC BLACK 1984 Measurement of the pyrophosphate content of plant tissues. Plant Physiol 75:862-864
26. SMYTH DA, M-X WU, CC BLACK 1984 Pyrophosphate and fructose 2,6-bisphosphate effects on glycolysis in pea seed extracts. Plant Physiol 76:316-332
27. TANNER GJ, L COPELAND, JF TURNER 1983 Subcellular localization of hexose kinase in pea stems: mitochondrial hexokinase. Plant Physiol 72:659-663
28. TURNER JF, DH TURNER 1980 The regulation of glycolysis and the pentose phosphate pathway. In DD Davies, ed, The Biochemistry of Plants. Academic Press, New York, Vol 2, pp 279-316
29. WATSON JD 1965 Molecular Biology of the Gene. Benjamin, New York, p 158

30. WONG W-JL, T AP REES 1971 Carbohydrate oxidation in stele and cortex isolated from roots of *Pisum sativum*. Biochim Biophys Acta 252:296-304
31. WOOD HG, WE O'BRIEN, G MICHAELS 1977 Properties of carboxytransphosphorylase; pyruvate, phosphate dikinase; pyrophosphate-phosphofructokinase; pyrophosphate-acetate kinase and their roles in the metabolism of inorganic pyrophosphate. Adv Enzymol 45: 85-155
32. WU M-X, DA SMYTH, CC BLACK 1983 Fructose 2,6-bisphosphate and the regulation of pyrophosphate-dependent phosphofructokinase activity in germinating pea seeds. Plant Physiol 73:188-191

In vivo Chlorophyll *a* Fluorescence Transients Associated with Changes in the CO_2 Content of the Gas-Phase

DAVID A. WALKER and MIRTA N. SIVAK

Lavorel and Etienne (14) in their comprehensive and authoritative review of "In vivo Chlorophyll Fluorescence" called fluorescence a "rich and ambiguous signal" which is "no longer a subject for the specialists alone." Suitably encouraged, we have been trying for some time to diminish the ambiguity of the relationship to photosynthetic carbon assimilation. This has led us, for example, to attempt multiple simultaneous measurements of several aspects of photosynthesis (18, 19, 21, 23) and has left us in little doubt (Fig. 1) that, at least in certain circumstances, there is a clear, reciprocal, but phase-shifted, relationship between chlorophyll a fluorescence and carbon assimilation. It has also filled us with enthusiasm for an aspect of fluorescence pioneered by Heber and Krause (7, 11), which is still largely unexploited and which, it seems to us, is not only brimming with information about fundamentals, but potentially capable of practical application. We refer to what, for want of a better term, might be called "gas transients."

MATERIALS AND METHODS

Fluorescence (23) and light-scattering (18) were measured as before. Gas changes were effected by computer- and electronically-operated valves.

RESULTS AND DISCUSSION

Most of the work on chlorophyll a fluorescence in vivo has focused on the changes in fluorescence emission following re-illumination after a period of darkness (for reviews, see 13-15, 17, 21). This is fluorescence

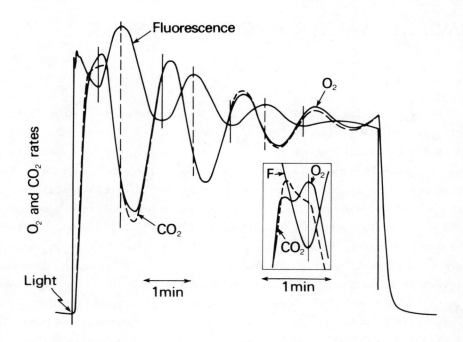

FIG. 1. The relationship between chlorophyll a fluorescence, CO_2 fixation, and O_2 evolution during oscillatory behavior by a spinach leaf. The leaf was photosynthesizing at 20°C in 2% CO_2 and 2% oxygen. The fluorescence signal maintains a broadly reciprocal relationship with gas exchange, but is phase-shifted (vertical bars are added to facilitate comparison) so that, e.g. a fluorescence peak anticipates a CO_2 and O_2 trough. Inset shows an example of the biphasic nature of the first gas-exchange peak, which is often observed under these conditions in which the principal O_2 peak then appears to coincide with the first fluorescence minimum.

induction, or the Kautsky (8) effect. It concerns the changes which occur as variable fluorescence parts company (at "O") from the constant background fluorescence and commences its initial rise ("I") through a dip ("D") to a peak ("P"), through a quasi-steady state (S), a secondary maximum ("M"), and finally falls to a true steady-state or terminal value ("T"). The terminology immediately conveys a sense of complexity but, in fact, it is still inadequate because the signals are often much more complex than this well-ordered sequence would suggest. Moreover, in passing from the fast events (O to P takes about a second and P to S an order of magnitude longer), the onset of carbon assimilation will come to influence fluorescence more and more strongly, and there is good evidence that "M peak," when it occurs, is associated with the termination of the transients at the induction period of carbon assimilation (21, 22).

The events which are responsible for induction in carbon assimilation are extremely complex and, for this reason, there is much to be said for studying changes in fluorescence occasioned by perturbations of the steady state other than re-illumination after a dark period. Gas changes are one of many ways in which the steady state may be perturbed (20) and, in relation to carbon assimilation, the air to CO_2-free air transient (Fig. 2) is simple and one which must, by definition, have a profound impact on carbon assimilation. We have confirmed (20) several of Krause's (11) and Heber's (7) original observations concerning the impact of CO_2 concentrations on fluorescence and have started to examine these in greater depth within the present conceptual framework of two main quenching mechanisms.

The first of these has been designated "q_Q" quenching. It derives from the idea of Duysens and Sweers (4) that a component of the photosynthetic electron transport chain ("Q" for "Quencher") initially oxidized, can accept electrons from chlorophyll a in PSII, thereby quenching the fluorescence which would constitute one channel of energy dissipation if Q were in the reduced state and unable to accept electrons. Q is now thought to comprise more than one type of plastoquinone, but the principle is the same. If Q is kept oxidized (by NADP and eventually by CO_2), it can accept electrons and quench fluorescence. According to this concept, if CO_2 is withdrawn, NADPH reoxidation should cease, and as Q becomes

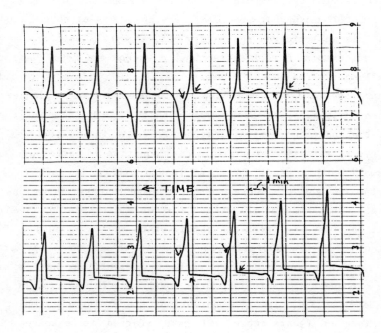

FIG. 2. Air to CO_2-free transients in wild-type and a photorespiratory mutant of barley. The upper trace shows the stability of the signal which is usually displayed (wild-type barley). It illustrates the initial rise following removal of CO_2 (downward arrow), the subsequent fall toward a lower steady-state value, and the reversal of this sequence, in essentials, following the reintroduction of CO_2 (upward arrow). The mutant, which dies at CO_2 levels normally found in air (ferredoxin-dependent glutamate synthase deficient, see Ref. 2) (lower trace), displayed a much less symmetrical transient, differing from the wild-type in several other details and declining in magnitude throughout the experiment.

reduced, q_Q quenching should relax and fluorescence should rise. Indeed, in some circumstances, this may account for some, if not all, of the initial rise in fluorescence which occurs during an air to CO_2-free air transient (Fig. 2 and 3A).

It will be seen, however, that this initial rise is quickly overtaken by a decline, and the final value in low or moderate light is lower rather than the higher value which would be demanded if this were the only mechanism involved. The secondary fall in fluorescence is now attributed to "q_e quenching." The mechanism remains uncertain, but it appears to be directly related to the proton gradient across the thylakoid membrane (12, for review see 13). When this is high, chlorophyll excitation energy, which would otherwise be dissipated as fluorescence, is dissipated as heat. Accordingly, when CO_2 fixation is halted, ATP will no longer be consumed in this process and ADP will, therefore, no longer be made available to discharge the proton gradient via the ATPase. Consequently, fluorescence will be quenched. If q_Q relaxes rapidly upon removal of CO_2 and q_e increases more slowly, but more extensively, these two mechanisms could account for the fluorescence transient which is often observed. Such a transient can be readily simulated by summing two exponential functions of opposite sign and appropriately different time-constants. The complete transient (i.e. air to CO_2-free air and back to air as in Fig. 2) is usually asymmetric. This can be explained by invoking the reverse of the responses outlined above. The asymmetry would then reside in the fact that when linear electron transport was resumed (on returning to air) it would do so against the higher "back-pressure" of the proton gradient built up (by cyclic and pseudo-cyclic electron transport) during the CO_2-free period.

Yet, as N. Baker (personal communication) has observed, a given fluorescence signal could be derived from more than one combination of quenching components and, at least, part of the initial rise (in Fig. 2) could arise from relaxation of q_e. He has based his conclusion on experiments using "light-doubling" in which brief periods of saturating light are added to the actinic light in order to drive Q fully reduced and thereby give a measure of its reduction status in the intervening periods of lower light (1, 16). We have arrived at similar conclusions using Heber's light-scattering technique (7). Light-scattering of a low-intensity green light beam by the leaf can be taken as an indicator of the thylakoid energization and has been shown to be a measure of the trans-thylakoid proton gradient and the adenylate status in chloroplasts and leaves (9, 10).

Figure 3 shows three examples of light-scattering and chlorophyll a fluorescence measured simultaneously (cf. 18 and 19) in one barley leaf. In the simplest case (Fig. 3A), light-scattering rises immediately as CO_2 is removed, and falls as it is restored. In short, it would seem that, in these circumstances at least, the initial rise must result from the relaxation of q_Q and the subsequent fall from the imposition of q_e. At higher CO_2 (900 µl/L rather than 350 µl/L), the fluorescence transient is essentially the same (Fig. 3B), but the light-scattering rise is preceded by a dip, and the light-scattering fall is preceded by a rise (i.e. relaxation of q_e must contribute to the initial rise in fluorescence). In the most complex case (Fig. 3C), the Pi concentration in the cytosol has been decreased by mannose feeding (see below and also Fig. 4) and the CO_2 concentration is still 900 µl/L. In these circumstances, most of the light-scattering change is negative even though it changes direction when the CO_2 is reintroduced. In addition, the interval between gas exchanges is dominated by the large increase in fluorescence which could continue to display itself as a series of dampening oscillations (18, 23) if it were not for the next gas exchange.

The dips in light-scattering could be explained as follows. If the proton gradient decreases as CO_2 is removed from the gas phase, it must do so because depletion of the gradient outpaces its replenishment. The principal mechanism for replenishment is linear electron transport from water to CO_2. When this can no longer occur, the proton gradient can be built up by linear transport to oxygen (or other acceptors), or by cyclic electron transport; both more likely to be slower processes. Yet, the proton gradient will also continue to be discharged by ADP. If metabolite pool sizes (particularly 3-phosphoglyceric acid, PGA) are high, ADP regeneration, and proton gradient discharge might be expected to continue relatively unabated for some time after the abrupt fall in the linear electron transport rate brought about by CO_2-removal. This view is

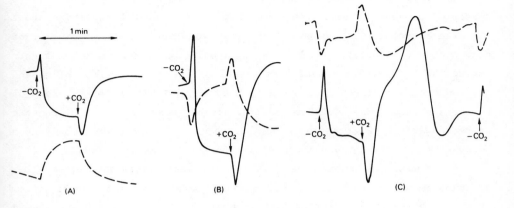

FIG. 3. Light-scattering and fluorescence transients induced by gas changes in barley. Presented here are examples of light-scattering signals (broken lines) associated with fluorescence transients (continuous lines) induced by gas changes in photosynthesizing barley (A) change from 350 µl/L CO_2 to zero and back. Note fairly typical (20) fluorescence transient and a light-scattering signal which rises simply as fluorescence falls and <u>vice versa</u>. B, Same leaf undergoing a gas change from 900 µl/L CO_2 to zero. Note exaggerated fluorescence transient (<u>cf</u>. A and B) and corresponding change in light-scattering, an indicator of the q_e component (C). The light-scattering signal displays both an initial dip and a corresponding rise as the CO_2 is replaced; same leaf undergoing gas exchange from 900 µl/L to zero after mannose feeding. The fluorescence transient is now complicated by the initiation, following the introduction of CO_2, of oscillations, the first of which dominates the entire transient. In these conditions, most of the light-scattering signals between the removal of CO_2 and its re-introduction are negative with respect to the starting point (<u>cf</u>. B which is mostly positive and A, which is all positive).

supported by the fact that in transients from relatively high CO_2 to CO_2-free air, an initial fall in light-scattering can be seen (cf. Fig. 3, A and B), and the proton gradient (as indicated by light-scattering) may not be restored to its initial value throughout much of the duration of a 30-s exposure to CO_2-free air (Fig. 3C).

Clearly, an enormous amount of work remains to be done before it is possible to amass a really adequate body of fundamental knowledge about the precise nature and mechanism of "gas transients" under a variety of conditions. Evidently, light-doubling could be combined with light-scattering to give more precise estimates of the contributions of q_Q and q_e, and these continuous measurements need to be linked with the assay of key metabolites if a complete picture is to emerge. As a diagnostic probe, however, gas transients may prove to be extremely useful even before the underlying fundamentals have been fully established.

As we have previously noted (21), a diagnostic probe should, ideally, be unambiguous, nonintrusive, convenient, sensitive, and robust (i.e. it should be relatively insensitive to what the investigator might regard as irrelevant factors). In some applications, it should give early warning of an impending change. Whether or not gas transients will prove to have all of these desirable features remains to be established. Certainly, they are relatively nonintrusive in the sense that intact leaves can be examined. The measurements themselves are simple, readily undertaken, and amenable to computerized analysis.

We have now started to assess the remaining criteria and wish to report two aspects which we find encouraging. The first concerns current work in collaboration with Peter Lea, which involves an examination of two barley photorespiratory mutants, one deficient in glutamate synthase, and the second, unable to convert glycine to serine (2). These mutants can grow normally at 0.5% CO_2, but are unable to grow in air. For the measurement of chlorophyll fluorescence during transients in the gas phase, leaves were illuminated at moderate light-intensity, and every 2 min, CO_2 was removed from the gas phase for 30 s (Fig. 2). After a few minutes of illumination, leaves of the wild-type strain showed reproducible transients on removal of CO_2, and leaves from different plants and ages showed a similar, recognizable pattern. Conversely, although leaves from the two mutant strains at first displayed kinetics not very different from those of the wild-type, the pattern (size and shape) changed continuously and soon became very different to the wild-type. This, of course, came as no surprise, because gas-exchange data indicated that this treatment would be deleterious for both mutants. Although this sort of analysis has the advantage of simplicity (no IRGA required) and is quick (a few minutes will show that the treatment is deleterious), no reproducibility can be obtained in the case of the mutants and comparison becomes difficult. Additional information could be obtained if the gas-transients were directly relevant to the genetic changes for which the investigator was screening. The two photorespiratory mutants studied in this case are defective in two different enzymes. Leaves of these strains will photosynthesize at rates comparable with the wild-type, if they are illuminated in low O_2 concentration. Under these conditions, the fluorescence kinetics were very different from one another and from the wild-type, if the O_2 concentration of the gas phase was increased, briefly, to 21 or 50%, and then decreased back to the starting level of 1%. Interpretation of these changes in chlorophyll fluorescence kinetics is specially complicated, because O_2 concentration affects not only photorespiration but also redox poising of the electron transport chain and pseudocyclic electron transport. However, the fact that these mutants showed distinctive patterns of fluorescence kinetics means that the use of these gas-transients for screening of genetic-manipulated plants is a real possibility. It is worth noting that these particular mutants can be recognized by gas-exchange techniques or simply by comparing growth at high CO_2 with growth at normal CO_2 concentration. However, other mutants might be detected by fluorescence measurements alone, during suitable gas changes.

The second aspect is an example of what we believe to be specific and indirect chemical intervention. Clearly, it would not be surprising if agents which interacted directly with the photochemical apparatus were to affect gas transients. We have, therefore, sought to examine the gas-transient response to mannose feeding, which is believed to exert its effect on photosynthesis by interacting in cytoplasmic events. Mannose is phosphorylated in the cytosol by glucose-6-P hexokinase but, in some species, mannose phosphate is not readily metabolized and the effect of feeding is then to sequester Pi in the cytosol (5). For this reason, mannose feeding affects photosynthesis by lowering the amount of Pi available for the phosphate translocator. The consequences are complex, but include lowering the rate of photosynthesis, increasing the incorporation of its products into starch, and abolishing the stimulation of photosynthesis brought about by lowering the O_2 concentration (6). Morrison and Batter (personal communication) have used L-mannose as a control in mannose feeding experiments and have reported it to be without effect, whereas the D isomer produced marked and rapid effects on carbon assimilation in wheat leaves. Figure 4 shows that D-mannose caused a 2- to 3-fold increase in the air to CO_2-free air transient in spinach whereas L-mannose was without effect. This response could be reversed by Pi feeding. Barley was particularly responsive, and Figure 5 shows the effect of mannose feeding on gas-transients in which the CO_2 concentration had been increased to 900 µl/L. In these and other experiments with spinach in which laboratory air (500-900 µl/L) was used, mannose feeding was associated with the initiation of oscillatory behavior.

Again, much more needs to be done before the nature of the response can be explained with confidence, but one aspect which is immediately apparent is the induction of oscillatory behavior. At this stage, we are unable to explain the nature of the mannose-induced changes in fluorescence which become apparent before there is any detectable effect on the rate of photosynthesis. It is clear, however (cf. 18, 23), that mannose feeding tends to induce oscillatory behavior (and Pi feeding tends to prevent oscillatory behavior) at CO_2 concentrations lower than those in which oscillation would normally be seen (and see also Ref. 6). Oscillations, we believe, are caused by an over-reaction of a regulatory mechanism, to perturbations of steady-state photosynthesis. According to this hypothesis (18, 23), the steady state is a balance between high ATP/ADP ratios (which favor the conversion of PGA to triose-phosphate) and low ATP/ADP ratios (which favor electron transport by facilitating the discharge, through the ATPase, of the proton gradient which would, otherwise, exert a back-pressure). Low Pi will affect this balance because it will tend to create low ATP/ADP ratios (as the rate of supply of Pi starts to limit

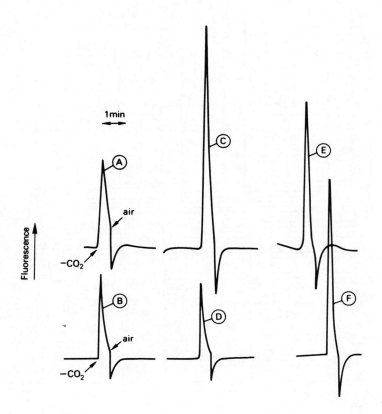

FIG. 4. Changes in the air to CO_2-free air fluorescence transients in spinach during mannose feeding. A, Signal from leaf at onset of feeding with D-mannose (20 mM). B, Signal from similar leaf at onset of feeding L-mannose. C and D, As for (A) and (B), but after 18-min feeding. E and F are as for (C) and (D), but for leaves fed with 50 mM Pi (E) or D-mannose (F) after a further 18-min feeding.

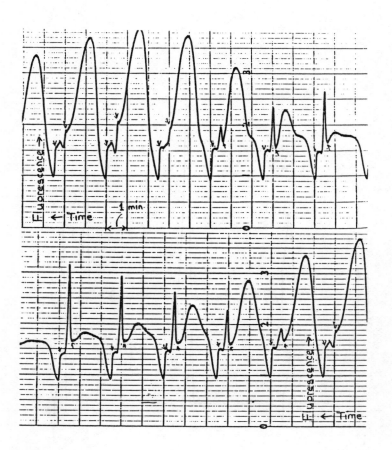

FIG. 5. Changes in gas-transients in barley leaves in high CO_2 air. Sequence starts at top right with addition of 20 mM D-mannose and progresses to bottom left, with ca. 900 µl/L CO_2 (see Fig. 2). Inorganic Pi (100 mM) fed for duration of three transients after completion of fifth transient. This effect of mannose feeding and its reversal by Pi could be repeated with the same leaf.

ATP formation) and increase metabolite pool sizes (as external Pi starts to limit triose-phosphate export). In addition, low ATP/ADP ratios will tend to be compensated by increases in PGA as the equilibrium of PGA/glycerate-1,3-bisphosphate is adjusted by mass action (3, 9). Pi will, therefore, act as a destabilizing factor just as high CO_2 and rapid rates of electron transport do. These destabilizing factors also tend to promote oscillations in response to perturbation of the steady state (22). Accordingly, when linear electron transport is interrupted in high light or low Pi by removal of CO_2, the proton gradient will continue to be discharged for longer times because of continuing regeneration of ADP (as ATP is consumed in the phosphorylation of PGA and ribulose-5-P). This would account for the dip in light-scattering which is sometimes observed (particularly in high light), and its exaggeration when Pi is sequestered by mannose feeding (Fig. 3).

CONCLUDING DISCUSSION

The changes in chlorophyll a fluorescence which accompany changes in the gas phase from air to CO_2-free air (and back again) are rapid, reproducible, and characteristic. Much of what is seen may be explained in terms of two main quenching mechanisms, q_Q and q_e, which are linked, respectively, to NADPH and ATP consumption in carbon assimilation. It is clear, however, from light-scatterng measurements, that there are circumstances in which these two mechanisms act in concert rather than in opposition to one another (see Fig. 3 and also Ref. 18).

While much remains to be done before these interrelationships can be fully defined, it seems that, at the purely practical level, the "gas transients" have potential as a diagnostic probe. Preliminary results suggest that the transients may be used to distinguish between mutant and wild-type leaves, and if such differences can be detected in untreated leaves, it clearly ought to be possible to demonstrate additional differences in appropriately designed feeding experiments. The present mannose feeding experiments were chosen to demonstrate that specific intervention in cytoplasmic events can also be rapidly reflected in gas-transient behavior. Clearly, this is the case, and with barley in particular, the gas-transient responses could be manipulated at will by alternating mannose feeding with Pi feeding. Before the nature of these changes can be properly explained, more work is needed, but the very fact that specific intervention in cytoplasmic events can bring about rapid and characteristic changes in fluorescence transients, adds to our belief that this type of diagnostic probe could play an increasingly useful role in the investigation of plant metabolism and its regulation.

LITERATURE CITED

1. BRADBURY M, NR BAKER 1981 Analysis of the slow phases of the in vivo chlorophyll induction curve. Changes in the redox state of photosystems I and II. Biochim Biophys Acta 635:542-551
2. BRIGHT SWJ, PJ LEA, P ARRUDA, NP HALL, AC KENDALL, AJ KEYS, JSH KUEH, ML PARKER, SE ROGNES, JC TURNER, RM WALLSGROVE, BJ MIFLIN 1984 Manipulation of key pathways in photorespiration and amino acid metabolism by mutation and selection. In PJ Lea, GR Stewart, eds, The genetic manipulation of plants and its application to agriculture. Oxford University Press, pp 141-169
3. DIETZ KJ, S NEIMANIS, U HEBER 1984 Rate limiting factors in leaf photosynthesis. II. Electron transport and ribulose-1,5-bisphosphate regeneration. Biochim Biophys Acta. In press
4. DUYSENS LNM, HE SWEERS 1963 Mechanism of two photochemical reactions in algae as studied by means of fluorescence. In Jap Soc Plant Physiol, ed, Studies on Microalgae and Photosynthetic Bacteria. University of Tokyo, Tokyo, pp 353-372
5. EDWARDS GE, DA WALKER 1983 C_3, C_4: Mechanisms and cellular and environmental regulation of photosynthesis. Blackwell Scientific Publications, Ltd, Oxford, pp 1-542
6. HARRIS GC, JK CHEESBROUGH, DA WALKER 1983 Measurement of CO_2 and H_2O vapour exchange in spinach leaf discs. Plant Physiol 71:102-107
7. HEBER U 1969 Conformational changes of chloroplasts induced by illumination of leaves in vivo. Biochim Biophys Acta 180:302-319
8. KAUTSKY H, A HIRSCH 1931 Neue Versuche zur Kohlenstoffassimilation. Naturwissenschafte 19:964

9. KOBAYASHI Y, S KOSTER, U HEBER 1982 Light-scattering, chlorophyll a fluorescence and state of the adenylate system in illuminated spinach leaves. Biochim Biophys Acta 682:44-54
10. KOSTER S, U HEBER 1982 Light-scattering and quenching of 9-aminoacridine fluorescence as indicators of the phosphorylation state of the adenylate system in intact spinach chloroplasts. Biochim Biophys Acta 680:88-94
11. KRAUSE GH 1973 The high energy state of the thylakoid system as indicated by chlorophyll fluorescence and by chloroplast shrinkage. Biochim Biophys Acta 292:715-728
12. KRAUSE GH, JM BRIANTAIS, C VERNOTTE 1981 Two mechanisms of reversible fluorescence quenching in chloroplasts. In G. Akoyunoglou, ed, Proc 5th Int Cong Photosynthesis, Vol I. Balaban Int Sci Serv, Philadelphia, pp 575-584
13. KRAUSE GH, E WEIS 1984 Chlorophyll fluorescence as a tool in plant physiology II. Interpretation of the fluorescence signals. Photosynth Res 5:139-157
14. LAVOREL J, AL ETIENNE 1977 In vivo chlorophyll fluorescence. In J Barber, ed, Primary Processes of Photosynthesis. Elsevier/North Holland Biomedical Press, Amsterdam, pp 203-268
15. PAPAGEORGIOU G 1975 Chlorophyll fluorescence: An intrinsic probe of photosynthesis. In Govindjee, ed, Bioenergetics of photosynthesis. Academic Press, New York, pp 319-371
16. QUICK WP, P HORTON 1984 Fluorescence induction in isolated barley protoplasts. Proc R Soc Lond B 220:361-382
17. SCHREIBER U 1983 Chlorophyll fluorescence yield changes as a tool in plant physiology: I. The measuring system. Photosynth Res 4:361-373
18. SIVAK MN, K-J DIETZ, U HEBER, DA WALKER 1984 The relationship between light-scattering and chlorophyll a fluorescence during oscillations in photosynthetic carbon assimilation. Arch Biochem Biophys. In press
19. SIVAK MN, U HEBER, DA WALKER 1984 Chlorophyll a fluorescence and light-scattering kinetics displayed by leaves during induction of photosynthesis. Planta. In press

20. SIVAK MN, RT PRINSLEY, DA WALKER 1983 Some effects of changes in the gas phase on the steady-state chlorophyll a fluorescence exhibited by illuminated leaves. Proc R Soc Lond B 217:393-404
21. SIVAK MN, DA WALKER 1984 Theory and practice of chlorophyll a fluorescence in its relation to photosynthesis: the state of the art and perspectives. <u>In</u> Proc OECD workshop on "Photosynthesis and physiology of the whole plant," September 24-26, Braunschweig, Germany
22. WALKER DA 1981 Secondary fluorescence kinetics of spinach leaves in relation to the onset of photosynthetic carbon assimilation. Planta 153:273-278
23. WALKER DA, MN SIVAK, RT PRINSLEY, JK CHEESBROUGH 1983 Simultaneous measurement of oscillations in oxygen evolution and chlorophyll a fluorescence in leaf pieces. Plant Physiol 73:542-549

Fine Control of Sucrose Synthesis by Fructose-2,6-Bisphosphate

MARK STITT

The aim of this article is to outline how fructose-2,6-bisphosphate (Fru-2,6-P_2) is involved in the fine control of sucrose and starch synthesis in leaves, to consider how fine control may interact with other regulatory mechanisms, and to speculate about what long-term perspectives this control provides for manipulating photosynthetic metabolism.

FUNCTION AND CONTROL STRATEGIES

The major end products of photosynthesis are sucrose and starch. Sucrose is exported to the remainder of the plant, while starch is accumulated in the leaf. During the night, starch is remobilized, either for respiration in the leaf, or for conversion to sucrose which is exported (e.g. see 8). In this way, starch supports metabolism in the leaf, and allows export of photosynthate to be spread over the whole diurnal period. This may be of benefit in providing a continual supply of photosynthate for sinks, or ensuring that the transport pathways are more efficiently used. Such functions involve the use of starch after it has accumulated and, ideally, starch accumulation during photosynthesis would reflect the demand for starch in the night. There is evidence that this occurs, so that, for example, a greater percentage of the available photosynthate is retained in leaves as starch when the total amount of photosynthate is reduced by decreasing the light intensity or shortening the photoperiod (1, 2, 24).

In contrast, biochemical studies of starch accumulation in isolated chloroplasts have lead to a view that starch is an "overflow" product, which is accumulated when demand for sucrose does not match the rates of CO_2 fixation (e.g. see 12), rather than being laid down so that a particular amount of carbohydrate is available in the leaf in the subsequent period of darkness.

Probably both viewpoints are valid. A full appreciation of the role of photosynthate partitioning will require considering how it may be directed toward long-term goals, but any such long-term strategy will interact with short-term alterations in the rate at which carbon can be fixed, as well as with the way in which photosynthetic metabolism is balanced by fine control. Due to the design of the pathways involved in photosynthesis, the production of sucrose and starch cannot be viewed in isolation from the operation of photosynthesis, but will closely interact with the operation of the Calvin cycle.

COORDINATION OF SUCROSE SYNTHESIS AND CARBON FIXATION

<u>The Investment Dilemma.</u> Isolated chloroplasts convert CO_2, Pi, and water to triose-phosphate (triose-P) (see Ref. 7 for more details). The triose-P is exported via the phosphate translocator, in strict counterexchange for Pi, which is needed to support further photosynthesis. If too little Pi is available in the medium, then phosphorylated intermediates accumulate in the stroma, and photosynthesis is inhibited by a depletion of stromal Pi. However, ribulose-1,5-phosphate (RuBP) must also be regenerated if CO_2 fixation is to continue, and this will be prevented if too much Pi is present in the medium and triose-P is withdrawn so rapidly that stromal metabolite pools are depleted. In a leaf, chloroplasts are not surrounded by a large volume of medium, acting as an "infinite" sink for triose-P and source of Pi. Instead, triose-P is removed to synthesize sucrose in the cytosol, regenerating Pi which can be recycled to the chloroplast. Maximal rates of photosynthesis will not be attained unless the rates of CO_2 fixation and sucrose synthesis are coordinated so that the fluxes are brought into balance at a point where the triose-P/Pi ratio does not restrict the rate of photosynthesis. In other words, the fine control of sucrose synthesis has to carry out a task which is analogous to that of an experimenter when he identifies the Pi optimum for isolated chloroplasts, and then mixes the assay

components together. How is this done, and is the correct answer always chosen?

Fructose-2,6-Bisphosphate Signals the Availability of Photosynthate.

The discovery of a regulatory metabolite called Fru-2,6-P_2 in liver (13) has led to a revision of our concepts about the control of glycolysis and gluconeogenesis. Fru-2,6-P_2 is present in the cytosol of leaves in µM concentrations (28) and modifies the activity of target enzymes like the cytosolic fructose-1,6-bisphosphate-1, phosphatase (FbPase) and the PPi-Fru-6-P phosphofructophosphotransferase (PFPase, 5, 14, 27). Fru-2,6-P_2 is a signal metabolite rather than an intermediate in the metabolic pathways, and is synthesized and degraded by specific enzyme activities termed fructose-6-phosphate, 2-kinase (Fru6P,2kinase) and Fru-2,6-P_2 phosphatase. Both of which are found in spinach leaves (3-5). The concentration of Fru-2,6-P_2 varies, depending on the relative activity of these enzymes, which are regulated by metabolites such as dihydroxyacetone-phosphate (DHAP), 3-phosphoglyceric acid (PGA), Pi, and Fru-6-P (3, 5, 29).

Fru-2,6-P_2 is involved in sensing how much photosynthate is available to make sucrose, and in adjusting the rate of sucrose synthesis accordingly. In Figure 1A, the rate of photosynthesis of spinach leaf discs was varied by using different light intensities or CO_2 concentrations. As photosynthesis rates rise, the amount of DHAP in the leaf material increases, while Fru-2,6-P_2 decreases (30). The higher DHAP reflects an increasing availability of photosynthate, but to understand how its use is controlled, we must look at the contribution of Fru-2,6-P_2. Fru-2,6-P_2 is a potent inhibitor of the cytosolic FbPase, acting at µM concentrations and inducing sigmoidal substrate dependence (see Refs. 5, 14, 27). The cytosolic FbPase occupies a strategic site in metabolism, catalyzing the first irreversible reaction during the conversion of triose-P to sucrose in the cytosol. This means that synthesis of sucrose from DHAP will be promoted by the lowered levels of Fru-2,6-P_2. In fact, the increased DHAP is at least partly responsible for the alteration of Fru-2,6-P_2 (and, hence, stimulates its own use) as it inhibits Fru6P,2kinase, the enzyme which synthesizes Fru-2,6-P_2 (29). Figure 1B compares the in vivo relation between DHAP and Fru-2,6-P_2 with the

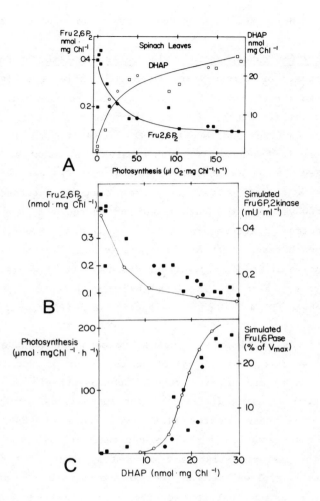

FIG. 1. Feedforward regulation of the cytosolic Fru-1,6-phosphatase in spinach leaves. A (top), Metabolite content when photosynthesis is varied by altering light (■, □) or CO_2 (●, ○). B (middle) and C (bottom), Comparison of simulated activity of enzyme (right axis, ○——○) with alterations of Fru-2,6-P_2 or photosynthesis (left axis, ●, ■). Fru6P,2kinase was assayed with 5 mM Pi, 2 mM Fru-6-P, and it is assumed DHAP is in the cytosol (volume, 20 µl · mg^{-1} Chl). Fructose-1,6-bis-phosphate-1, phosphatase (FbPase) is abbreviated as Fru1,6Pase above.

simulated effect of DHAP on the activity of Fru6P,2kinase, making the assumptions that the DHAP is only in the cytosol (with a volume of 20 µl mg^{-1} Chl, so 20 nmol DHAP mg^{-1} Chl is represented by 1 mM DHAP in the assay). The decreased Fru-2,6-P_2 could be accounted for by the sensitivity of Fru6P,2kinase to inhibition by DHAP.

DHAP not only stimulates the cytosolic FbPase by decreasing the concentration of inhibitor, but will also increase the substrate concentration, as DHAP is converted to Fru-1,6-P_2 by reactions which are near to equilibrium. The response of the cytosolic FbPase to DHAP can be modeled (14) using theoretical equilibrium constants to predict the relation between DHAP and Fru-1,6-P_2, and the results from Figure 1B to predict the empirical relation between DHAP and Fru-2,6-P_2 in leaves. The isolated cytosolic FbPase activity can then be assayed in the presence of a predicted Fru-1,6-P_2:Fru-2,6-P_2 ratio. Figure 1C predicts how the activity of the cytosolic FbPase (solid line) should vary, as DHAP concentration increases, and compares this with the measured rates of photosynthesis in leaf discs containing different amounts of DHAP (solid symbols). The agreement between the observed rate of photosynthesis and the predicted activity of the cytosolic FbPase provides evidence that this control mechanism is an important element in coordinating sucrose synthesis with photosynthesis in leaves. Yet, the quantitative aspects should not be overinterpreted because of the assumptions made in our model.

Up to a threshold value, the cytosolic FbPase is effectively inactive. It may be supposed that at concentrations of DHAP below this threshold, there are inadequate levels of metabolites in the stroma for turnover of the Calvin cycle, and so the plant cannot "afford" to make sucrose. However, above this threshold, there is a strong activation of the FbPase in response to small increments of DHAP. Evidently, there are now adequate levels of metabolites to allow turnover of the Calvin cycle and the "surplus" may be removed to make sucrose. The sensitivity of this activation may be important to ensure that sucrose synthesis is stimulated to remove the DHAP and regenerate Pi before an accumulation of phosphorylated intermediates start to inhibit photosynthesis.

This model will certainly require modification and elaboration. For example, our current work is already showing how cytosolic metabolites interact with Fru-2,6-P_2 in regulating the cytosolic FbPase. Regulation of sucrose-phosphate (sucrose-P) synthase by the Glc6P:Pi ratio (6) is also important in ensuring that hexose-phosphate (hexose-P) produced by FbPase is converted to sucrose without large alterations in metabolite concentrations (see Ref. 30). However, to test the validity of the concept that the cytosolic FbPase is regulated to ensure that adequate concentrations of metabolites are maintained to allow photosynthesis, we have looked at a plant where a specialized metabolism requires higher concentrations of DHAP to be maintained during photosynthesis.

<u>Intercellular Shuttles in Maize Require a Higher "Threshold" for Sucrose Synthesis</u>. In maize, regeneration of RuBP requires a rapid intercellular transport of metabolites, including DHAP, as well as turnover of the Calvin cycle. The Calvin cycle is restricted to the bundle-sheath chloroplasts, but these are deficient in photosystem II and cannot reduce all the PGA produced by carboxylation of RuBP (7, 10); instead about half of the PGA moves out into the mesophyll cells, where it is reduced to DHAP before returning to the bundle-sheath chloroplasts. Hatch and Osmond (10) suggested this intercellular transport might occur by diffusion through the symplasm, but pointed out that concentration gradients of 10 mM would be needed between the two cell types. Recently, two different methods for studying the intercellular distribution of metabolites in maize leaves have been developed (21, 32), and both confirm that there are large concentration gradients for PGA (high in the bundle sheath, low in the mesophyll) and for DHAP (high in the mesophyll, low in the bundle sheath). During photosynthesis, there is over 10 times more DHAP in the mesophyll cells of maize leaves than in spinach leaves (Table I; see also Refs. 21, 32). However, recent work has also shown that most of the sucrose synthesis in maize leaves occurs in the mesophyll cells (9).

This poses a dilemma, for how can removal of DHAP for sucrose synthesis be reconciled with the maintenance of the high concentrations of DHAP in the mesophyll which are needed to drive intercellular diffusion. An enzyme like the cytosolic FbPase from spinach leaves, which apparently operates so that about 1 mM DHAP is present during photosynthesis, would undermine the diffusion gradients needed in maize. The results in Table I suggest that the maize mesophyll cytosolic FbPase does not become active

Table I. Relation Between Fructose-2,6-Bisphosphate, Dihydroxyacetone Phosphate, and Photosynthesis in Spinach and Maize

The leaves were taken from plants which had been predarkened for 15 h, and the treatment was for 10 min (spinach) or 20 min (maize). For more details, see Refs. 30 and 32.

	$Fru-2,6-P_2$		DHAP	
	Spinach	Maize	Spinach	Maize
	($nmol \cdot mg^{-1}$ Chl)			
Light, air	0.08	0.07	22	440
Light, no CO_2	0.19	0.21	14	160
Dark, air	0.18	0.19	3	8

until substantial concentrations of DHAP are available; maize leaves illuminated in the absence of CO_2 degraded their sucrose (not shown), but still contained 160 nmol DHAP mg^{-1} Chl, which is 6-fold higher than is found in spinach leaves during rapid sucrose synthesis. This difference between maize and spinach does not seem to be due to maize containing much higher levels of $Fru-2,6-P_2$ than spinach (Table I), but reflects an alteration in the properties of the FbPase. The maize mesophyll cytosolic FbPase requires 7- to 15-fold higher concentrations of substrate than the enzymes from spinach or wheat leaves (Table II; 14, 27, 32).

This adaptation of the "tuning" of the fine control mechanisms in maize raises the question whether other plants may have adapted these control systems to their own particular metabolic "problems." It is also of interest that the major change seems to have been in the properties of a target enzyme (FbPase) rather than in the levels of $Fru-2,6-P_2$. Altering the target enzyme may be a more specific way of adapting a given regulatory response than altering $Fru-2,6-P_2$, which probably controls many other aspects of carbohydrate turnover.

Table II. Comparison of the Substrate Affinity of the Cytosolic Fructose-1,6-Bisphosphate-1, Phosphatase in Spinach, Wheat, and Maize

The enzyme was assayed at pH 7.1 with 4 mM $MgCl_2$, 1.3 mM EDTA. For further details, see Refs. 15, 27, and 32.

Additions in assay	$Fru-1,6-P_2$ concentration (µM) needed for half-maximal activity		
	Spinach	Wheat	Maize
None	3	4	20
1 µM $Fru-2,6-P_2$	55	42	500
10 µM $Fru-2,6-P_2$	250	180	3500

FINE CONTROL OF PARTITIONING

Involvement of $Fru-2,6-P_2$. So far, we have considered how the cytosol senses how much carbon the chloroplasts can provide for sucrose synthesis without impairing their own operation. However, information also returns to the chloroplast and signals how much demand there is for further synthesis of sucrose. This can be seen in leaves where photosynthesis is faster than the rate of sucrose export, so that sucrose starts to accumulate in the leaf. Frequently, a point is reached where sucrose accumulation slows down, or stops, and the "surplus" carbohydrate is diverted into starch in the chloroplast (8, 24, 26).

$Fru-2,6-P_2$ is involved in this "feedback" control of sucrose synthesis (Fig. 2). Figure 2A shows how $Fru-2,6-P_2$ increases when sucrose accumulates in spinach leaf discs or maize segments carrying out rapid photosynthesis. In these experiments, the sucrose content was increased by illuminating detached leaf material for varying lengths of time, but similar results have been obtained by varying the demand for sucrose in the whole plants, or by adding exogenous sugars to discs (31, unpublished results for maize). The higher $Fru-2,6-P_2$ concentration restricts the activity of the cytosolic FbPase, so that more photosynthate is retained in the chloroplast for synthesis of starch (28, 31). This is illustrated

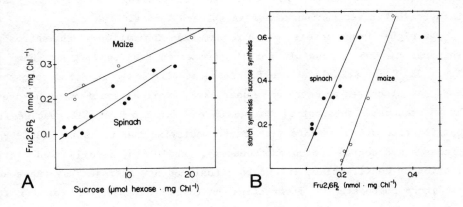

FIG. 2. Involvement of fructose-2,6-bisphosphate in photosynthate partitioning. A, Relation between sucrose content of leaf material and Fru-2,6-P_2 content. B, Influence of Fru-2,6-P_2 on partitioning into starch and sucrose.

in the experiment of Figure 2B, where spinach leaf discs or maize leaf segments were allowed to photosynthesize for varying lengths of time to allow them to accumulate differing amounts of sucrose, before measuring the rate of sucrose and sucrose synthesis, and the level of Fru-2,6-P_2. In both maize and spinach, there is a strong positive correlation between increased Fru-2,6-P_2 and increased partitioning of photosynthate into starch.

Interaction of Fru-2,6-P_2 and Sucrose-P Synthase. The evidence available at present suggests that sugars do not directly modify Fru-2,6-P_2 concentration, as they have no direct effect on the activity of Fru6P,2kinase or Fru-2,6-P_2-phosphatase (28). However, the accumulation of Glc6P (Fig. 3A), Fru6P, and UDP-glucose (data not shown) shows that the activity of sucrose-P synthase is restricted when sucrose accumulates (see also 23, 26). It is not yet clear how this happens. Sucrose does not inhibit spinach leaf sucrose-P synthase (15). However, Huber and co-workers have shown that the activity of sucrose-P synthase in extracts from soybean leaves varies diurnally (26), possibly due to diurnal rhythms (19, 26), and also changes within hours when the demand for sucrose is altered (25). We do not yet know what is causing these changes, and whether similar alterations occur in spinach leaves.

The higher Fru6P levels, which result from inhibition of the sucrose-P synthase, may be responsible for the increased Fru-2,6-P_2. As shown in Figure 3B, Fru6P stimulates Fru6P,2kinase and inhibits Fru-2,6-P_2 phosphatase. As discussed by Rufty and Huber (25), small alterations of hexose-P should produce a substantial change in the Fru-2,6-P_2 level, because of amplification which results from Fru6P modifying the rates of Fru-2,6-P_2 synthesis and degradation simultaneously. While many details need clarification or confirmation, it can be visualized how sucrose-P synthase and Fru-2,6-P_2 cooperate in restricting sucrose synthesis when demand for sucrose decreases.

Response of Chloroplast Metabolism. A restriction of sucrose synthesis would lead to lower rates of photosynthesis unless control mechanisms are available which can direct the "surplus" photosynthate into other end products before they accumulate to a point where low Pi begins to limit the rate of photosynthesis. There is a sensitive regulation of chloroplast

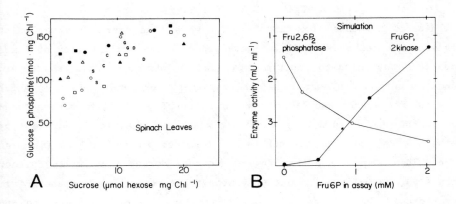

FIG. 3. Relation between hexose-P and fructose-2,6-bisphosphate. A, Increased Fru-2,6-P_2 in response to higher glucose-6-P in spinach leaves. B, Regulation of spinach leaf Fru-6-P,2kinase and Fru-2,6-P_2 phosphatase from spinach leaves by Fru6P.

metabolism in leaves. As $Fru-2,6-P_2$ increases with increasing sucrose concentration (Fig. 2A), there is only a small increase of DHAP (Fig. 4B), but a very sharp increase in the rate of starch synthesis (Fig. 4B). This resembles the marked stimulation of starch synthesis in isolated chloroplasts when the external Pi is decreased (12) or the triose-P/Pi ratio is increased (18).

This stimulation of starch synthesis involves an interaction between PGA reduction and ADP-glucose pyrophosphorylase, the enzyme catalyzing the first irreversible reaction leading to starch. The ADP-glucose pyrophosphorylase is stimulated (23) as the PGA/Pi ratio increases in isolated chloroplasts under conditions with low Pi (12, 22). The PGA/Pi ratio will be a sensitive indicator for a decreased supply of Pi in the stroma, because low Pi typically leads to a high PGA/triose-P ratio (12) as the photogeneration of ATP or NADPH is restricted. However, since PGA reduction is reversible and can be driven at high rates by high ATP and NADPH or by high levels of PGA (11), a stimulation of starch synthesis does not necessarily need to be accompanied by a decreased rate of photosynthesis.

INTERACTION OF REGULATORY SYSTEMS

These experiments show how leaves have a fine control system which recognizes and responds to: (a) fluctuations in the availability of fixed carbon for conversion to end products, and (b) changes in the short-term demand for sucrose. Thus, rapid photosynthesis produces triose-P, which lowers the $Fru-2,6-P_2$ and favors sucrose synthesis, but if sucrose accumulates in the leaf, $Fru-2,6-P_2$ increases again and restricts sucrose synthesis, so that more photosynthate is retained in the chloroplasts as starch. This sequence of events can be seen during the day in spinach plants (28), and allows the use of photosynthate to respond to short-term alterations in the balance between the availability of photosynthate and the demand for sucrose. While doing this, the fine control system is directed toward a "goal" of maintaining metabolic conditions in the leaf which allow effective use of the available photosynthetic machinery.

However, distribution of photosynthate will also depend upon the relative amounts of different enzymes in the leaf. For example, the

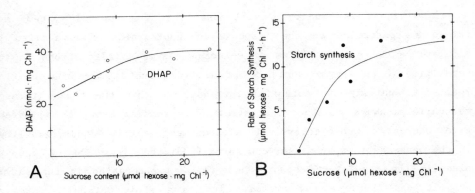

FIG. 4. Relation between sucrose accumulation, increased dihydroxyacetone phosphate content, and starch accumulation. A, Sucrose and DHAP. B, Sucrose and starch synthesis. (From same experiments as in Fig. 2.)

activity of some Calvin cycle enzymes and starch synthesizing enzymes may vary, depending on the photoperiod (24), and the activity of sucrose-P synthase is varied depending on genetic components (17), on photoperiod, nitrogen source, developmental stage (16, 19), CO_2 concentration and water stress (18). Although definitive evidence is lacking, at least some of these changes may reflect an alteration in the amount of protein available. Such alterations of capacity will interact with fine control systems.

In some cases, it can be envisaged that alteration of capacity will cooperate with fine control in producing the change of flux. For example, any change of sucrose-P synthase capacity should lead to alterations of $Fru-2,6-P_2$, which then modify conditions in the chloroplast to achieve a diversion of photosynthate from sucrose to starch (see above). Another example of such cooperation was recently described (24). Spinach plants grown in short days partition more of their photosynthate into starch, and also have higher photosynthetic rates. The higher photosynthesis reflected increased capacities of some Calvin cycle enzymes, rather than of electron transport, and was consequently associated with higher levels of PGA. This increased PGA would stimulate ADP-glucose pyrophosphorylase (23) and could be partly responsible for diverting more photosynthate toward starch.

However, it is equally possible that the fine control systems may act as a constraint on the response of photosynthetic metabolism to alterations in the capacities of selected enzymes. Alterations in the capacities of an enzyme would be of little value if the fine control reacted to nullify the potential increase, so that the fluxes through metabolic pathways remained unaltered. This implies that, to be effective, the enzyme whose activity is altered must exhibit "controllability" in the sense defined by Kacser and Burns (20): the enzyme should occupy a place in the regulatory network such that an alteration in the enzyme activity can impose a new overall flux on a whole metabolic pathway. Many enzymes which are loosely referred to as "regulatory" may not be capable of dominating a pathway in this way because fine control mechanisms damp out any increment in flux which is introduced at the enzyme. Identification of enzymes with "controllability" will be necessary if changes in the activity or properties of a given enzyme are to be used to modify metabolic fluxes or performance. For example, the general correlation between sucrose-P synthase activity and partitioning into starch suggests that this enzyme is capable of determining fluxes in leaves (15-19). Likewise, the kinetic properties of the cytosolic FbPase--its substrate affinity and sensitivity to inhibitors--may be crucial in controlling metabolite and Pi concentrations during photosynthesis.

Another possibility which arises is that alterations of capacity may be detrimental, if they introduce such a shift in metabolism that fine control mechanisms are no longer capable of maintaining the "poise" between different reactions or metabolite pools which is required to allow rapid photosynthesis. In maize, for example, a modification of the fine control of sucrose synthesis was needed to accommodate the metabolite concentrations needed for photosynthesis in these specialized plants. This is obviously more extreme than the changes which will occur within a given species, but we do not know how tightly fine control imposes a constraint on metabolic pathways--how soon would alterations in the capacity of an enzyme start to exhaust the ability of the fine control systems to maintain conditions which allow these changes to be exploited. For example, the results presented in this article predict that there will be fluctuations in the concentrations of Pi, triose-P, and other metabolites when a leaf adjusts the rates of starch and sucrose synthesis to short-term changes in conditions, the rate of photosynthesis, or accumulation of sucrose, as well as to long-term changes in the capacities of enzymes like sucrose-P synthase. But how sensitive and flexible are the regulatory systems? Are the fluctuations of metabolites so small that they never limit the operation of the Calvin cycle? In this case, there may be considerable freedom to manipulate the partitioning of photosynthate between sucrose and starch, before the rate of photosynthate is impaired. On the other hand, if metabolite concentrations often fluctuate outside the "band" which allows optimal operation of the Calvin cycle, then adaptation of this fine regulation of sucrose (or starch) may be needed to realize fully the photosynthetic potential in varying conditions or when partitioning is being altered.

Acknowledgments—Many of the results described here were a result of collaborative work with R. Gerhardt, B. Herzog, and H. W. Heldt in Göttingen, as well as with Ch. Foyer and R. Furbank (Sheffield) and A. Balogh, C. Cseke, and B. B. Buchanan (Berkeley). Financial support was provided by the Deutsche Forschungsgemeinschaft, and from the National Science Foundation during visits to Berkeley.

LITERATURE CITED

1. CHATTERTON NJ, JE SILVIUS 1979 Photosynthetic partitioning into starch in soybean leaves. I Effect of photoperiod versus photosynthetic period duration. Plant Physiol 64:749-753
2. CHATTERTON NJ, JE SILVIUS 1972 Photosynthate partitioning into starch as affected by daily photosynthetic period duration in six species. Plant Physiol 49:141-144
3. CSEKE C, BB BUCHANAN 1983 An enzyme synthesizing fructose2,6bisphosphate occurs in leaves and is regulated by metabolite effectors. FEBS Letts 155:139-142
4. CSEKE CM, M STITT, A BALOGH, BB BUCHANAN 1983 A product-regulated fructose 2,6 bisphosphatase occurs in green leaves. FEBS Lett 162:103-106
5. CSEKE C, N WEENDEN, BB BUCHANAN, K UYEDA 1982 A special fructosebisphosphate functions as a regulatory metabolite in green leaves. Proc Natl Acad Sci USA 79:4322-4326
6. DOEHLERT DC, SC HUBER 1984 Phosphate inhibition of spinach leaf sucrose phosphate synthase as affected by glucose 6 phosphate and phosphoglucoisomerase. Plant Physiol 76:250-253
7. EDWARDS GE, DA WALKER 1983 C_3C_4: mechanisms, and cellular and environmental regulations of photosynthesis. Blackwell, London
8. FONDY BR, D GEIGER 1982 Diurnal rhythm of translocation and carbohydrate metabolism in Beta vulgaris L. Plant Physiol 70:671-676
9. FURBANK B, C FOYER, M STITT Intercellular compartmentation of sucrose synthesis in leaves of Zea mays. Planta. In press

10. HATCH MD, CB OSMOND 1976 Compartmentation and transport in C_4 photosynthesis. In CR Stocking, U Heber, eds, Encyclopedia of Plant Physiol, Vol 3, New Series. Springer-Verlag, Berlin, pp 144-148
11. HEBER U 1984 Flexibility of chloroplast metabolism. In C Sybesma, ed, Advances in Photosynthetic Research. Proc. VI International Congress on Photosynthesis, Vol 3. Nijhoff-Junk, The Hague, pp 381-389
12. HELDT HW, CJ CHON, D MARONDE, A HEROLD, ZS STANKOVIC, DA WALKER, A KRAMINER, MR KIRK, U HEBER 1977 Role of orthophosphate and other factors in the regulation of starch formation in leaves and isolated chloroplasts. Plant Physiol 59:1146-1155
13. HERS H-G, L HUE, E van SCHAFTINGEN 1982 Fructose2,6bisphosphate. Trends Biochem Sci 7:329-331
14. HERZOG B, M STITT, HW HELDT 1984 Control of photosynthetic sucrose synthesis by fructose2,6bisphosphate III. Properties of the cytosolic fructose-1,6-bisphosphatase. Plant Physiol 75:561-565
15. HUBER SC 1981 Interspecific variation in activities and regulation of leaf sucrose phosphate synthase. Z Pflanzenphysiol 102:443-450
16. HUBER SC, DW ISRAEL 1982 Biochemical basis for partitioning of photosynthetically fixed carbon between starch and sucrose in Soybean (Glycine max Merr.) leaves. Plant Physiol 69:691-696
17. HUBER SC 1983 Role of sucrose phosphate synthase in partitioning of carbon in leaves. Plant Physiol 71:818-821
18. HUBER SC, HH ROGERS, FL MOWRY 1984 Effects of water stress on photosynthesis and carbon partitioning in Soybean (Glycine max (L.) Merr.) plants grown in the field at different CO_2 levels. Plant Physiol 76:244-249
19. HUBER SC, TW RUFTY, PS KERR 1984 Effect of photoperiod on photosynthate partitioning and diurnal rhythms in sucrose phosphate synthase activity in leaves of soybean (Glycine max L. (Merr)) and Tobacco (Nicotinia tabacum L). Plant Physiol 75:1080-1084
20. KACSER H, JA BURNS 1973 A Theory of Metabolic Regulation. Symp Soc Exp Biol 27:65-104
21. LEEGOOD RC The intercellular compartmentation of metabolites in leaves of Zea mays. Planta. In press

22. PORTIS AR 1982 Effects of the relative concentrations of inorganic phosphate, 3-phosphoglycerate and dihydroxyacetone phosphate on the rate of starch synthesis in isolated chloroplasts. Plant Physiol 70:393-396
23. PREISS J 1982 Regulation of the biosynthesis and degradation of starch. Annu Rev Plant Physiol 33:431-454
24. ROBINSON JM 1984 Photosynthetic carbon metabolism in leaves and isolated chloroplasts from spinach plants grown under short and intermediate photosynthetic periods. Plant Physiol 75:397-409
25. RUFTY TW JR, SC HUBER 1983 Changes in starch formation and activities of sucrose phosphate synthase and cytoplasmic fructose1,6bisphosphatase in response to source-sink alterations. Plant Physiol 72: 474-480
26. RUFTY TW JR, PS KERR, SC HUBER 1983 Characterization of diurnal changes in activities of enzymes involved in sucrose biosynthesis. Plant Physiol 73:428-433
27. STITT M, G MIESKES, H-D SÖLING, HW HELDT 1982 On a possible role of fructose2,6bisphosphate in regulating photosynthetic metabolism in leaves. FEBS Lett 145:217-222
28. STITT M, R GERHARDT, B KÜRZEL, HW HELDT 1983 A role for fructose-2,6-bisphosphate in the regulation of sucrose synthesis in spinach leaves. Plant Physiol 72:1139-1141
29. STITT M, C CSEKE, BB BUCHANAN 1984 Regulation of fructose2,6bisphosphate concentration in spinach leaves. Eur J Biochem 143:89-93
30. STITT M, B HERZOG, HW HELDT 1984 Control of photosynthetic sucrose synthesis by fructose2,6bisphosphate. I. Coordination of CO_2 fixation and sucrose synthesis. Plant Physiol 75:548-553
31. STITT M, B KÜRZEL, HW HELDT 1984 Control of photosynthetic sucrose synthesis by fructose2,6bisphosphate. II. Partitioning between sucrose and starch. Plant Physiol 75:554-560
32. STITT M, HW HELDT Control of photosynthetic sucrose synthesis by fructose2,6bisphosphate. IV. Intercellular metabolite distribution and properties of the cytosolic fructosebisphosphatase in maize leaves. Planta. In press

Compartmentation of Glycolytic Enzymes in Plant Cells

DAVID T. DENNIS, WILMA E. HEKMAN, ALAN THOMSON, ROBERT J. IRELAND, FREDERIK C. BOTHA, and NICHOLAS J. KRUGER

Plastids play an essential role in the biosynthetic activity of both photosynthetic and nonphotosynthetic plant cells. For example, fatty acid biosynthesis occurs exclusively in plastids (27) as does the biosynthesis of the majority of amino acids (26). Since each tissue will have its own specific requirement for intermediates, the complement of enzymes within plastids may vary. It is assumed that in most cases, plastid enzymes are encoded in the nuclear genome, translated in the cytosol, and imported into the plastid. The mechanism by which the synthesis and import of enzymes into plastids is regulated is as yet unknown. However, this regulation is likely to be under nuclear control.

In the case where identical reactions occur in the cytosol and plastid, the reactions are catalyzed by isozymes (7). The genes for each pair of isozymes are probably located in the nuclear genome, but this has been shown for only a small number of the isozymes that have been characterized. The proportions of cytosolic and plastid isozymes are dependent on the particular tissue and on the stage of development of that tissue. For example, in the developing endosperm of castor oil seed, 70% of the total cellular phosphogluconate dehydrogenase is located in the leucoplast, whereas only 10% is in the plastids of germinating seeds (28, 29). Obviously, a major clue to the control of differentiation will be the determination of the mechanism by which these ratios are regulated. The leucoplasts in developing endosperm of castor oil seed have been studied as a model system to investigate the differentiation of plastids (7).

LEUCOPLASTS FROM CASTOR OIL SEEDS

All types of plastid are probably derived from a common precursor termed a proplastid (19, 34). During development, this proplastid is thought to differentiate into the specific type of organelle that is characteristic of the tissue in which it is located. The metabolism of leucoplasts in the developing endosperm of castor oil seeds is directed primarily toward fatty acid biosynthesis (7). These organelles appear to have lost the capacity to differentiate into chloroplasts since they will not become green, even on prolonged exposure to light. The leucoplasts vary considerably in shape and have a dense matrix (Fig. 1). Early in development they may contain starch grains, but these tend to be lost later.

Little is known about the similarities of DNA from different types of plastid. In the case of chloroplasts from the bundle-sheath and mesophyll cells of C_4 plants, which have different enzyme complements and metabolic functions, the DNA molecules are identical (35), indicating that the difference in their metabolism is not due to changes in the DNA during development. However, DNA from plastids with radically different functions such as the endosperm leucoplasts and leaf chloroplasts of the castor oil plant have not been studied.

It is possible that there is heterogeneity of the DNA in proplastids and that proplastids with different DNAs give rise to the variety of plastids that are found in various tissues. Alternatively, plastid DNA may be modified during development by, for example, excision or methylation of specific segments. To test these hypotheses, DNA has been extracted and characterized from leaf chloroplasts and endosperm leucoplasts.

LEUCOPLAST AND CHLOROPLAST DNA

Details of the methods for the extraction and purification of leucoplast DNA will be published elsewhere. Both leucoplast and chloroplast DNAs are circular molecules (Fig. 2). The contour length of leucoplast DNA, as measured on the electron microscope, is 136.6 $\pm$ 0.9 kbp, and is identical with that of chloroplast DNA (137.6 $\pm$ 2.0 kbp) within the limits of accuracy of the method. Restriction endonuclease digestion of leucoplast and chloroplast DNA with either PstI, XhoI, BglII or HindIII

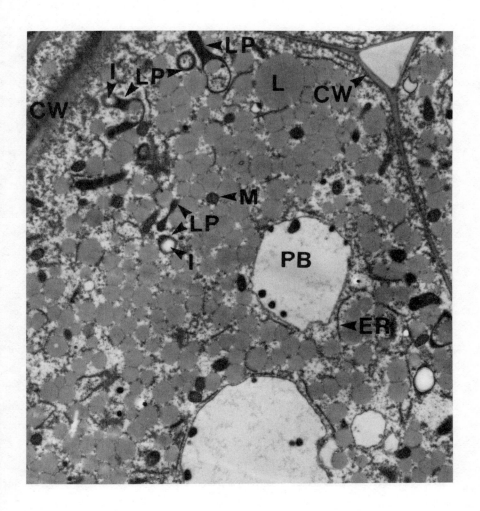

FIG. 1. Electron micrograph of cell from developing castor oil seed endosperm. With a magnification of 5,320 X, the picture shows leucoplasts (LP), inclusion bodies (I), lipid bodies (L), protein bodies (PB), endoplasmic reticulum (ER), mitochondria (M), and cell wall (CW).

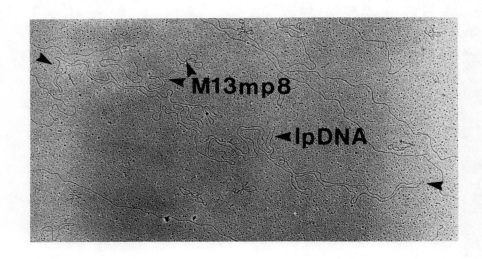

FIG. 2. Electron micrograph of DNA extracted from leucoplasts of the endosperm of developing castor oil seed. Magnification of 13,850 X. The two arrows mark the extremities of a leucoplast DNA molecule. Bacteriophage M13mp8 DNA was included as a standard.

produced identical pairs of restriction patterns. Similarly, identical restriction fragments were obtained on digestion of both these DNAs with the isoschizomeric enzymes HpaII and MspI, only one of which will cut when the bases are methylated. This result indicates that neither DNA is methylated at these sites. As a final test, Southern blots of restriction endonuclease digestions of the two DNAs were probed with cloned BamHI fragments from castor oil seed leucoplast and Vicia faba chloroplast DNAs (the latter kindly provided by N. Straus, Toronto University). Individually, both of these probes gave identical patterns with the two DNAs.

The above data indicate that the gross differences in morphology and enzyme components that exist between the leucoplast from the endosperm and the chloroplast from the leaf of the castor oil plant are not due to differences in the DNA of the organelles. Assuming that most, if not all, of the glycolytic enzymes are encoded in the nuclear genome, the origin of the differences must be sought in the coding sequence and promoter regions of the nuclear genes coding for the leucoplast and cytosolic isozymes of the glycolytic pathway. As a first step in investigating this process, the glycolytic isozymes have been isolated, purified, and characterized.

LEUCOPLAST AND CYTOSOLIC ISOZYMES OF THE GLYCOLYTIC PATHWAY

The reactions of the leucoplast glycolytic pathway are catalyzed by enzymes that are isozymes of their cytosolic counterparts, and from which they can be separated by ion exchange chromatography (24). In some cases, the cytosolic and leucoplast isozymes have almost identical physical and kinetic properties, whereas in other instances the isozymes are quite different.

Previously, we have proposed that the leucoplast isozymes may be important in the synthesis of fatty acids. Sucrose, entering the endosperm from the leaves, is converted to glucose and fructose by invertase which is located exclusively in the cytosol (30). The hexoses may then enter the leucoplast where they can be metabolized by the leucoplast glycolytic pathway to pyruvate and subsequently to acetyl-CoA and fatty acids. A scheme for this pathway, and the evidence for it, has been presented elsewhere (7). However, subsequent studies suggest two modifications to this scheme. First, uptake of glucose-6-phosphate is unlikely to occur (25). Second, leucoplasts are able to metabolize external 3-phosphoglycerate, possibly due to the presence of a phosphate translocator (25). Therefore, triose-phosphates (triose-P) may provide an alternative source of substrate for leucoplast fatty acid synthesis. At the moment, the importance in vivo of the two entry points into the pathway of fatty acid biosynthesis is not known.

The properties of the glycolytic isozymes from castor oil seeds have been reviewed previously (7). Therefore, in the remainder of this section, we shall concentrate primarily on more recent work on selected enzymes rather than a description of the entire pathway.

Hexose-Phosphate Isomerase. Two isozymes of hexose-phosphate isomerase from the endosperm can be separated by ion exchange chromatography (Fig. 3). The cytosolic isozyme elutes at a lower conductivity than the leucoplast isozyme, and is approximately twice as active. The isozymes have identical native mol wt of 130 to 140,000, as determined by gel filtration chromatography and PAGE using the Ferguson method (12). SDS-PAGE showed that both isozymes are composed of two identical subunits of mol wt 66 to 70,000. Similarly, the substrate kinetics with regard to fructose-6-phosphate are very similar for both the isozymes with K_m values of 0.33 and 0.24 mM for the cytosolic and leucoplast isozymes, respectively. The isozymes are active over a wide range of pH with identical optima of 8.2. Hence, the only apparent difference between the isozymes is their difference in charge. Similar results have been found for the isozymes for Clarkia in which, although the subunits had similar physical and kinetic properties, a mixture of the dissociated subunits would only form homodimers when reassociated; plastid-cytosolic heterodimers did not occur (36). Hence, in both castor oil seed and Clarkia, the very similar properties of the isozymes may conceal more fundamental differences in the structure of the proteins.

Phosphofructokinase. The function of the phosphofructokinases in developing castor oil seed endosperm is now uncertain since this tissue has both ATP- and PPi-dependent phosphofructokinases (PFKase and PFPase, respectively). The latter enzyme is present in many plant tissues (3) and is activated by Fru-2,6-P_2 (20). In the endosperm, the activity of

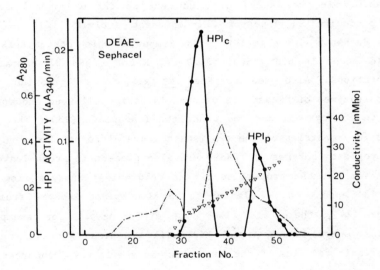

FIG. 3. Separation of the plastid and cytosolic forms of hexose-phosphate isomerase on DEAE-Sephacel. ▼, conductivity; •, enzyme activity; -·-, protein; HPI, hexose-phosphate isomerase; c, cytosolic; p, plastid.

PFPase, which is entirely cytosolic, is 5 to 10 times that of the ATP-dependent enzyme when assayed under optimal conditions in the presence of Fru-2,6-P_2. The importance of PFPase in glycolysis in this tissue is not known. A glycolytic pathway using PFPase has been shown to operate in extracts of other tissues that are supplied with PPi (32), and PPi has been found in sufficient quantities for PFPase to be active (11, 31).

In spinach leaves, PFKase and PFPase are reported to be interconvertable (1, 37), but this is not the case for the enzymes from developing endosperm. In the latter tissue, the enzymes are antigenically distinct proteins. Antibodies raised against leucoplast PFKase are capable of immunoprecipitating both PFKase isozymes, but do not affect PFPase. Similarly, antibodies against PFPase do not affect PFKase (Fig. 4). The apparent interconversion of these two enzymes reported by others (37) may, in part, be explained by contamination of the coupling enzymes by UDP-glucose pyrophosphorylase. In the presence of PPi, this contaminant generates UTP in the assay which can act as the phosphoryl donor for PFKase, producing an apparent PPi-dependent activity (21). (Editors' note: B. Buchanan sent a letter to the Symposium which stated this sort of error did occur in his research. Thus, Refs. 1 and 37 contain wrong interpretations. Also see ap Rees *et al.*, this Symposium p. 78).

Two isozymes of PFKase are present in castor oil seed endosperm; one located in the cytosol and the other the leucoplast (13). The physical and kinetic properties of the two isozymes are different (14). Cytosolic and chloroplast isozymes of PFKase are also present in green leaves (17). Although some of the properties of the chloroplast isozyme from spinach leaves (18) appear to be similar to the leucoplast isozyme from castor oil seeds (14), there are also some differences. For example, the chloroplast enzyme shows normal Michaelis kinetics, whereas one of the characteristic features of the leucoplast enzyme is its pronounced sigmoid kinetics. Previously, we have reported that the leucoplast isozyme was activated by Fru-2,6-P_2 (23), however, we have been unable to repeat this result using commercial Fru-2,6-P_2.

Ion exchange chromatography of an extract from leaves of castor oil plant produces two peaks of PFKase (Fig. 5). These peaks elute at exactly the same conductivities as the isozymes in extracts of the endosperm.

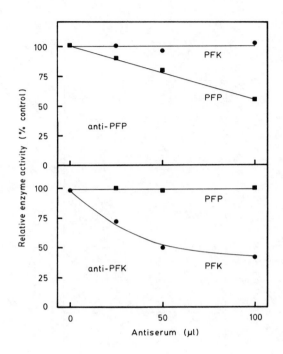

FIG. 4. Immunoprecipitation of ATP-dependent phosphofructokinase and PPi-dependent phosphofructophosphotransferase. Aliquots of a desalted, crude homogenate of castor oil seed endosperm were incubated with rabbit antiserum raised against either potato tuber PFPase (upper panel, labeled PFP) or endosperm leucoplast PFKase (lower panel, labeled PFK) for 1 h. Each sample was then incubated for 1 h with IgGsorb and centrifuged. The supernatant was assayed for both PFKase (●) and PFPase (■). Each value is the mean of two determinations. The results are expressed as a percentage of the initial activities which were 9.7 and 112.5 nmol · min^{-1} · ml^{-1} for the PFKase and PFPase, respectively. Preimmune serum did not precipitate either activity.

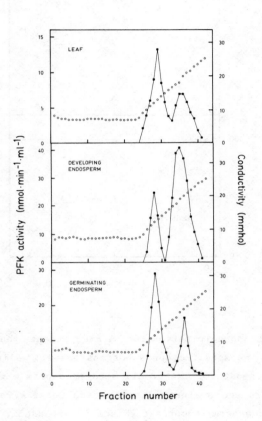

FIG. 5. DEAE-Sephacel chromatography of ATP-dependent phosphofructokinase isozymes from castor oil plants extracted from: upper panel, leaves; middle panel, developing endosperm; lower panel, germinating endosperm. The recovery was at least 88% in each case. The fractionation was previously described (13). PFK = PFKase.

However, preliminary results indicate that the chloroplast enzyme is not identical to the leucoplast enzyme. It is possible that the leucoplast and chloroplast enzymes represent separate proteins or, alternatively, they may be post-translationally modified.

Our knowledge of the properties of plant PFKase is still fragmentary. In many reports, isozymes have not been considered, and it is not possible to deduce which isozyme has been studied. It is also probable that the enzyme may have properties that are characteristic of the tissue in which it is located. When combined with the observation that at least some, if not the bulk of, cytosolic glycolysis may proceed via PFPase, these considerations indicate that our knowledge of the control of this important step of glycolysis in plants is rudimentary.

NAD Glyceraldehyde-3-Phosphate Dehydrogenase. The NAD-specific glyceraldehyde-3-phosphate dehydrogenase has been purified to homogeneity from pea seeds and the kinetic mechanism determined (8, 9). This is now accepted as the mechanism for the enzyme from all sources studied (10). Plants also contain an NADP-specific enzyme that is completely distinct from the NAD-specific enzyme with different subunits and possibly a different kinetic mechanism (4-6).

Isozymes of the NAD-specific glyceraldehyde-3-phosphate dehydrogenase have been separated from castor oil seed endosperm, the leucoplast enzyme accounting for ca. 14% of the total activity (7). Isozymes of this enzyme also occur in developing pea seed, but in this case, they occur in equal amounts (Hekman and Dennis, unpublished). It is not clear, as yet, which of the two pea isozymes is present in the amyloplast and which is cytosolic. The dried pea seed contains only one isozyme.

The two isozymes from the castor oil seed endosperm have been partially purified and their properties determined. The pH optima of the two isozymes are essentially identical. However, the kinetic properties of the isozymes are different. The $\underline{K_m}$ values of the leucoplast enzyme for both NAD^+ and Pi are considerably higher than those of the cytosolic form, whereas in the case of glyceraldehyde-3-phosphate, the $\underline{K_m}$ of the cytosolic form is higher. In addition, the cytosolic enzyme is quite unstable in the absence of NAD^+. The activity of the inactive enzyme can be restored by incubation with NAD^+, although it takes several hours for the restoration of full activity. The leucoplast enzyme, in contrast, is stable in the absence of NAD^+, and the addition of NAD^+ does not increase activity.

Antibodies have been made to the homogeneous pea seed enzyme. These antibodies have been shown by Western blot analysis to cross-react only with one of the isozymes of the developing pea seed. In contrast, the antibodies cross-react with both isozymes from castor oil seed. The significance of this result is not yet known.

Phosphoglyceromutase. Phosphoglyceromutase has been reported to be present in leucoplasts from the castor oil seed endosperm (30) and cauliflower florets (15). Endosperm leucoplasts that have been purified by washing and centrifugation through sucrose contain the enzyme, but no cytosolic contamination as judged by the absence of alcohol dehydrogenase. By comparison with the distribution of RuBP carboxylase (Rubisco), we estimate that leucoplasts contain 5 to 7% of the total endosperm phosphoglyceromutase. Phosphoglyceromutase has been reported to be absent from chloroplasts, and a shuttle of 3-phosphoglycerate out of the chloroplast and 2-phosphoglycerate into the chloroplast has been proposed to bypass this step (33). The absence of the enzyme in chloroplasts would prevent the loss of carbon from the Calvin cycle (33). In contrast, the presence of the enzyme in plastids that are not involved in photosynthesis would facilitate the flow of carbon from hexoses to such products as fatty acids that are synthesized in the plastids.

The isozymes of phosphoglyceromutase are very difficult to separate, although a separation by ion filtration chromatography has been reported (24). Phosphoglyceromutase has been purified to homogeneity by ammonium sulfate precipitation, hydrophobic column chromatography, DEAE-Sephacel chromatography, hydroxyapatite chromatography, and both nondenaturing and denaturing PAGE. The details of the purification and properties of this enzyme will be presented elsewhere (2). Because of the difficulty of separating the isozymes, the purified protein probably contains both isozymes; however, the preparation is apparently free from other proteins.

Antibodies raised against the purified protein have been used to measure the amount of this enzyme during castor oil seed development. The activity of both isozymes of phosphoglyceromutase increases greatly during seed development, but falls again as the seed matures; the dry seed having virtually no activity. There is a further rise and fall of the activity in the endosperm during germination. However, when the phosphoglyceromutase protein is measured by ELISA using the antibody raised against the purified protein, the level of the protein appears to remain high and constant in the dried seed. This apparent contradiction is explained by Western blots of total extracts of the endosperm at different stages of development. These show that when the enzyme activity is low in the dried seed, the ELISA technique is detecting fragments of the protein. Hence, the enzyme is at least partially broken down as the seed dries and new protein is synthesized during germination.

The individual isozymes of phosphoglyceromutase have been partially purified (as far as the hydroxyapatite stage described above) by separating the leucoplasts and purifying the enzyme in the supernatant and in the purified leucoplasts. The isozymes have identical mol wt of $63,000 \pm 7,000$ as determined by gel filtration on HPLC. The other properties of the isozymes are also very similar, but differences do occur. The leucoplast isozyme is considerably more thermolabile than the cytosolic isozyme. Both isozymes have activity over a wide pH range in the forward and reverse directions; however, the optima are slightly, but significantly, different. Similarly, the K_m values for the substrate in the forward and reverse direction are significantly different. The K_m values for 3-phosphoglycerate are 430 ± 40 and 330 ± 25 μM for the leucoplast and cytosolic isozymes, respectively. The corresponding K_m values for 2-phosphoglycerate are 112 ± 20 and 60 ± 10 μM.

Neither isozyme requires a divalent metal ion for activity, nor were any compounds found that regulate the activity of the isozymes. The animal enzyme requires 2,3-bisphosphoglycerate for activity and this compound has been reported to activate the leucoplast isozyme from cauliflower (15). However, 2,3-bisphosphoglycerate had no effect on either castor oil seed isozyme. An initial observation indicating a requirement for this metabolite may have been due to contamination of the coupling enzymes by animal phosphoglyceromutase.

Hence, the castor oil seed endosperm appears to have isozymes of phosphoglyceromutase which have similar but some distintly different properties. The similarity of the properties and the problems involved in separating the isozymes make it difficult to detect the presence of this enzyme in plastids from other tissues where the activity may be lower. However, the availability of antibodies may make this possible in the future.

DISCUSSION

It is clear that the endosperm of the developing castor oil seed has distinct leucoplast and cytosolic isozymes for each step in the glycolytic pathway. With some enzymes, such as phosphofructokinase and glyceraldehyde-3-phosphate dehydrogenase, the properties of the two isozymes are obviously different, whereas with others, for example hexose-phosphate isomerase and phosphoglyceromutase, the differences are more subtle. However, even for these latter enzymes, when the isozymes are examined in more detail, differences in both kinetic and physical properties become apparent. For the few enzymes that have been investigated, the isozymes are encoded on distinct nuclear genes. The difference in properties of each isozyme pair suggest that this organization is quite general. However, such differences could conceivably result from post-translational modifications. In addition, although location of the genes for some of the leucoplast isozymes on the leucoplast genome is unlikely, this possibility cannot be ruled out at present.

The complement of enzymes found in plastids appears to be highly variable and determined by the tissue in which the plastid is found. The difference in enzyme complement appears not be the result of changes in the DNA found in the plastid since, as described above, the DNA extracted from leaf chloroplasts and endosperm leucoplasts appears identical, even though these plastids have quite distinct functions. Therefore, the difference is almost certainly under nuclear control.

Plastids have distinct functions in different tissues, some of which are shown in Figure 6. In the developing oil seed, the main function of the plastid is to synthesize fatty acids. Carbon may enter the plastid either as hexose or as triose-P via the phosphate translocator. As expected, these plastids contain all of the enzymes for the conversion of hexose to fatty acid. In addition, the reverse pathway from

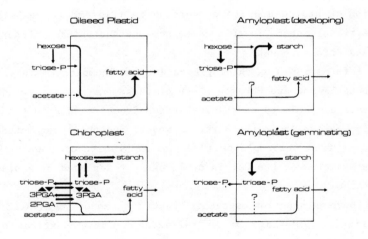

FIG. 6. Diagrammatic representation of the major pathways that can occur in various types of plastids.

triose-P to hexose, and hence to starch, is restricted in the leucoplast because of the absence of fructose-1,6-bisphosphate-1,phosphatase (FbPase) which is only present in the cytosol (J. A. Miernyk, personal communication).

In starch-storing seeds such as pea seed, amyloplasts are likely to contain all the enzymes required for the conversion of triose-P to hexose. ap Rees and co-workers (22) have demonstrated that these enzymes are present in amyloplasts from soybean tissue culture. They have suggested that cytosolic hexose is first converted to triose-P, which then enters the amyloplast via the phosphate translocator. However, direct incorporation of hexose into starch without the conversion to triose has also been reported (16). Amyloplasts from developing pea seed appear not to contain phosphofructokinase, phosphoglyceromutase or enolase (D. T. Dennis, unpublished results). The apparent absence of these glycolytic enzymes in such amyloplasts may account for the low levels of storage lipids in this seed.

As yet, there is no information on the enzyme complement of amyloplasts from starch-storing seeds during germination. However, it should be quite different since these plastids presumably will be involved in the degradation of starch and its conversion to triose-P for export via the phosphate translocator. In this case, phosphofructokinase (PFKase) may be present and FbPase absent.

The information for chloroplasts is more complete. These plastids contain both PFKase and FbPase. Both of these enzymes are required since chloroplasts are able to synthesize starch from triose-P in the light and conversely degrade it in the dark. However, phosphoglyceromutase appears to be absent from chloroplasts (33). A possible consequence of this is that intermediates of the Calvin cycle are not depleted by a plastid glycolytic pathway.

The above discussion suggests that the major metabolic function of a particular plastid type is reflected in its constituent enzymes. Undoubtedly, the enzyme complement of the plastids within a particular tissue is a major factor in the control of metabolism in that tissue. This is clearly illustrated not only by the enzymes that are found in the leucoplasts in the endosperm of the castor oil seed, but also the difference in the enzymes found in bundle-sheath and mesophyll chloroplasts of C$_4$ plants. As yet, we have little idea about how the enzyme complement of such plastids is regulated. However, such information is essential if the differentiation of various plastid types is to be understood, and possibly manipulated.

LITERATURE CITED

1. BALOGH A, JH WONG, C WOTZEL, J SOLL, C CSÉKE, BB BUCHANAN 1984 Metabolite-mediated catalyst conversion of PFK and PFP: A mechanism of enzyme regulation in green plants. FEBS Lett 169:287-292
2. BOTHA FC, DT DENNIS 1985 Phosphoglyceromutase from the developing endosperm of *Ricinus communis*. Isolation and characterization. Arch Biochem Biophys (submitted)
3. CARNAL NW, CC BLACK 1983 Phosphofructokinase activities in photosynthetic organisms - the occurrence of pyrophosphate-dependent 6-phosphofructokinase in plants and algae. Plant Physiol 71: 150-155
4. CERFF R 1978 Glyceraldehyde-3-phosphate dehydrogenase (NADP) from *Sinapis alba*: steady state kinetics. Phytochem 17:2061-2068
5. CERFF R 1982 Evolutionary divergence of chloroplast and cytosolic glyceraldehyde-3-phosphate dehydrogenases from angiosperms. Eur J Biochem 126:513-515
6. CERFF R, SE CHAMBERS 1979 Subunit structure of higher plant glyceraldehyde-3-phosphate dehydrogenase. J Biol Chem 254:6094-6098
7. DENNIS DT, JA MIERNYK 1982 Compartmentation of nonphotosynthetic carbohydrate metabolism. Annu Rev Plant Physiol 33:27-50
8. DUGGLEBY RG, DT DENNIS 1974 Nicotinamide adenine dinucleotide-specific glyceraldehyde 3-phosphate dehydrogenase from *Pisum sativum*. Purificiation and characterization. J Biol Chem 249: 162-166
9. DUGGLEBY RG, DT DENNIS 1974 Nicotinamide adenine dinucleotide-specific glyceraldehyde 3-phosphate dehydrogenase from *Pisum sativum*. Assay and steady state kinetics. J Biol Chem 249: 167-174

10. DUGGLEBY RG, DT DENNIS 1982 Glyceraldehyde-3-phosphate dehydrogenase from pea seeds. Methods Enzymol 89:319-325
11. EDWARDS J, T AP REES, PM WILSON, S MORRELL 1984 Measurement of the inorganic pyrophosphate in tissues of *Pisum sativum* L. Planta 162:188-191
12. FERGUSON KA 1964 Starch gel electrophoresis-application to the classification of pituitary proteins and polypeptides. Metabolism 13:985-1002
13. GARLAND WJ, DT DENNIS 1980 Plastid and cytosolic phosphofructokinases from the developing endosperm of *Ricinus communis* L. Separation, purification and initial characterization of the isozymes. Arch Biochem Biophys 204:302-309
14. GARLAND WJ, DT DENNIS 1980 Plastid and cytosolic phosphofructokinases from the developing endosperm of *Ricinus communis* II. Comparison of the kinetics and regulatory properties of the isoenzymes. Arch Biochem Biophys 204:310-317
15. JOURNET E-P, R DOUCE 1984 Capacity of cauliflower bud plastids to synthesize acetyl-CoA from 3-phosphoglycerate. Comptes Rendus des Seances de l'academie des Sciences 298:365-370
16. KEELING PL, RH TYSON, IG BRIDGES 1983 Evidence for the involvement of triose phosphates in the pathway of starch biosynthesis in developing wheat grain. Biochem Soc Trans 11:791-792
17. KELLY GJ, E LATZKO 1977 Chloroplast phosphofructokinase. I. Proof of phosphofructokinase activity in chloroplasts. Plant Physiol 60: 290-294
18. KELLY GJ, E LATZKO 1977 Chloroplast phosphofructokinase. II. Partial purification, kinetic and regulatory properties. Plant Physiol 60:295-299
19. KIRK JTO, RAE TILNEY-BASSETT 1978 The plastids. Their chemistry, structure, growth and inheritance, Rev 2nd ed. pp 219-241. Elsevier/North Holland Biomed, New York: 960 pp
20. KOMBRINK E, NJ KRUGER, H BEEVERS 1984 Kinetic properties of pyrophosphate:fructose-6-phosphate phosphotransferase from germinating castor bean endosperm. Plant Physiol 74:395-401

21. KRUGER NJ, DT DENNIS 1985 A source of apparent pyrophosphate: fructose 6-phosphate phosphotransferase activity in rabbit muscle phosphofructokinase. Biochem Biophys Res Commun 126:320-326
22. MACDONALD FD, T AP REES 1983 Enzymic properties of amyloplasts from suspension cultures of soybean. Biochim Biophys Acta 727:81-89
23. MIERNYK JA, DT DENNIS 1982 Activation of the plastid isozyme of phosphofructokinase from developing endosperm of Ricinus communis by fructose 2,6-bisphosphate. Biochem Biophys Res Commun 105: 793-798
24. MIERNYK JA, DT DENNIS 1982 Isozymes of the glycolytic enzymes in endosperm from developing castor oil seeds. Plant Physiol 69: 825-828
25. MIERNYK JA, DT DENNIS 1983 The incorporation of glycolytic intermediates into lipids by plastids isolated from the developing endosperm of castor oil seeds (Ricinus communis L.). J Exp Bot 34:712-718
26. MIFLIN BJ, PJ LEA 1977 Amino acid metabolism. Annu Rev Plant Physiol 28:299-329
27. OHLROGGE JB, DN KUHN, PK STUMPF 1979 Subcellular localization of acyl carrier protein in leaf protoplasts of Spinacia oleracea. Proc Natl Acad Sci USA 76:1194-1198
28. SIMCOX PD, DT DENNIS 1978 Isoenzymes of the glycolytic and pentose phosphate pathways in the proplastids from the developing endosperm of Ricinus communis L. Plant Physiol 61:871-877
29. SIMCOX PD, DT DENNIS 1978 6-phosphogluconate dehydrogenase isoenzymes from the developing endosperm of Ricinus communis L. Plant Physiol 62:287-290
30. SIMCOX PD, EE REID, DT CANVIN, DT DENNIS 1977 Enzymes of the glycolytic and pentose phosphate pathways in plastids from the developing endosperm of Ricinus communis L. Plant Physiol 59:1128-1132
31. SMYTH DA, CC BLACK 1984 Measurement of the pyrophosphate content of plant tissues. Plant Physiol 75:862-864
32. SMYTH DA, M-X WU, CC BLACK 1984 Pyrophosphate and fructose 2,6-bisphosphate effects on glycolysis in pea seed extracts. Plant Physiol 76:316-320

33. STITT M, T AP REES 1979 Capacities of pea chloroplasts to catalyze the oxidative pentose phosphate pathway and glycolysis. Phytochem 18:1905-1912
34. THOMSON WW, JM WHATLEY 1980 Development of nongreen plastids. Annu Rev Plant Physiol 31:375-394
35. WALBOT V 1977 The dimorphic chloroplasts of the C_4 plant <u>Panicum maximum</u> contain identical genomes. Cell 11:729-737
36. WEEDEN NF 1983 Evolutionary affinities of plant phosphoglucose isomerase and fructose 1,6-bisphosphatase isozymes. Isozymes: Curr Top Biol Med Res 8:53-66
37. WONG JH, A BALOGH, BB BUCHANAN 1984 Pyrophosphate functions as phosphoryl donor with UDP-glucose-treated mammalian phosphofructokinase. Biochem Biophys Res Comm 121:842-847

Metabolite Levels, Chloroplast Envelope Transport, and Chloroplast Metabolism

ARCHIE R. PORTIS, JR., PATRICIA G. RAY, and WILLIAM E. BELKNAP

In any given metabolic pathway, the concentrations of various intermediates provide a "picture" of the flux through the pathway. These concentrations provide a rapid means of coordinating the fluxes through the various partial reactions of the pathway, such that they become equivalent under steady-state conditions. With given concentrations of the input and output intermediates, most pathways will spontaneously reach a steady-state condition due to this internal coordination.

Therefore, it is of considerable interest to determine how the concentrations of intermediates of photosynthetic metabolism relate to the various possible photosynthetic fluxes. We currently have only limited information on how levels of various intermediates change with changing rates of photosynthesis and assimilate partitioning. Such information would considerably expand our current understanding of the regulation of photosynthesis.

The major means of metabolite communication between the stroma and the cytoplasm is the phosphate translocator (6), since the chloroplast uses light energy, converted to NADPH and ATP, to convert CO_2, phosphate, and water to triose-phosphate (triose-P). The major metabolites translocated by the phosphate translocator are phosphate (Pi), 3-phosphoglycerate (PGA), and dihydroxyacetone phosphate (DHAP). Therefore, levels of these metabolites might be a primary means for the coordination of chloroplastic and cytoplasmic metabolism. From this perspective, it is not surprising that starch synthesis is regulated by the stromal PGA:Pi ratio (19), and sucrose synthesis is regulated by DHAP:Pi at the fructose-1,6-bisphosphate-1,phosphatase (FbPase) step (4, 14, 20) and by Pi at sucrose-phosphate synthase (1, 5, 9). However, the question can be raised as to how the levels of these metabolites in one compartment (i.e. stromal) are related to the levels in the other (i.e. cytoplasmic), and how metabolism and the properties of the phosphate translocator influence this relationship. Furthermore, studies of the relationships between isolated chloroplast metabolism and the relative levels of these intermediates in the external medium can provide information on what the relative levels of these metabolites should be in the cytoplasm in vivo.

METHODS

Growth of spinach, isolation of chloroplasts, assays of CO_2 fixation, and starch synthesis were performed as previously described (17). O_2 evolution and CO_2 fixation were assayed simultaneously with an O_2 electrode and by removal of aliquots for determination of acid-stable counts. Details of the computer modeling of the phosphate translocator have been published elsewhere (18).

RESULTS

Relationships Between Stromal and External Metabolites as Determined by the Phosphate Translocator. We utilized a kinetic model for translocator-mediated diffusion (8) to analyze how metabolic reactions occurring in the stroma and the external concentrations of Pi, PGA, and DHAP (which in principle, would be influenced by cytoplasmic metabolism) influence the stromal concentrations of these metabolites (17). Typical results are shown in Figure 1. In this case, the external concentrations of Pi, PGA, and DHAP were 1.0, 1.5, and 1.5 mM, respectively, and the pH of the stromal compartment was 8.0, while that of the external space was 7.6. These concentrations are certainly within typical ranges measured recently in leaves (12). If we could ignore all stromal metabolism, the stromal concentrations (in mM) would be 1.64, 2.58, and 5.77 for Pi, DHAP, and PGA, respectively, assuming a total concentration for these metabolites of 10 mM. Thus, the external ratio of 1.0:1.5:1.5 is changed to a stromal ratio of 1.0:1.57:3.52 by the properties of the translocator.

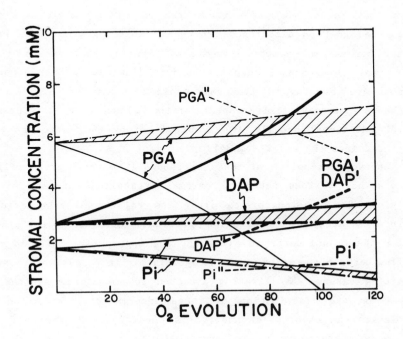

FIG. 1. Calculated effect of different types of carbon metabolism on the stromal concentations of Pi, 3-phosphoglyceric acid, and dihydroxyacetone phosphate. External concentrations of 1.0 mM for Pi, 1.5 mM for PGA, and 1.5 mM for DHAP were assumed. For more details, see Ref. 18. Hatched area represents CO_2 fixation exclusively with DHAP, listed as DAP (-), or PGA (o-o-) formation as boundaries. The effect of PGA reduction alone (——) or in combination with a fixed rate of CO_2 fixation (----, PGA', DAP', + Pi') and the use of external DHAP, listed as DAP, for ribulose-1,5-bisphosphate (RuBP) regeneration (---, PGA", DAP", + Pi") are also shown. The data have been calculated from a previous reference (17).

This effect is due to the fact that the translocator only appears to transport the doubly ionized forms of these compounds (6), and the pKa's of these three metabolites differ considerably. The actual ratio inside depends on the pH of the two compartments but, in general, a more alkaline stroma favors accumulation of PGA and, to a lesser extent, DHAP. It should be noted that recent work (7) with a reconstituted translocator system has indicated that pH differences between compartments may have effects on either the maximal velocity or the affinity of the translocator for certain types of exchange, such that under illumination DHAP export and PGA import are more favored than they would be otherwise. These effects would change some quantitative aspects of the calculated results presented here, but additional details are required before they actually can be incorporated into the kinetic model.

Of course, metabolism of these compounds is occurring, and Figure 1 illustrates how various rates of stromal metabolism interact with the capacity of the translocator to alter the relative concentrations of these intermediates. Concentration gradients must be established in order for the translocator to allow for the appropriate rates of metabolite flux between the stroma and the cytoplasm. It is easiest to examine the limiting cases. CO_2 fixation directed toward starch synthesis has no effect on the distribution of these intermediates since no fixed carbon leaves the chloroplast under this condition. Fixed CO_2 directed toward sucrose synthesis can theoretically be exported as either DHAP or PGA. As shown in Figure 1, stromal Pi declines with increasing flux (measured as O_2 evolution, necessary to establish Pi import) and either PGA or DHAP increase, depending on whether PGA or DHAP is the major export product. It is important to note that while the capacity of the translocator is very high (500 μmol mg^{-1} Chl h^{-1}, assumed for this case), it actually is not operationally higher, since the concentration of stromal Pi decreases rather rapidly and CO_2 fixation is theoretically limited to about 145 μmol O_2 mg^{-1} Chl h^{-1} with the parameters used for this example. Thus, high rates of CO_2 fixation favor increased starch formation, but can be limited by an inadequate capacity for metabolite exchange, causing stromal Pi to reach rate-limiting levels.

Furthermore, we have ignored the fact that high rates of flux into sucrose most certainly would be accompanied by decreasing cytoplasmic Pi levels.

External PGA may be reduced to DHAP, which effectively transfers photosynthetic energy to the cytoplasm, or external DHAP can be utilized for regeneration of RuBP with the result that CO_2 fixation only requires ATP. As shown in Figure 1, PGA reduction requires a rapid decrease in the stromal PGA concentration and a corresponding increase in the DHAP concentration, with only a slight change in Pi. Rates of reduction in excess of 100 μmol O_2 mg^{-1} Chl h^{-1} would not be possible with only PGA reduction. It should be noted that PGA reduction lowers the potential for starch synthesis, since a decrease in the stromal PGA:Pi ratios must occur if the external concentrations do not change. Conversely, the short circuiting of the normal reductive pentose phosphate cycle by importing external DHAP for RuBP regeneration lowers stromal DHAP, increases PGA, causes a slight increase in stromal Pi, and thereby favors starch synthesis.

The model clearly illustrates how changes in the relative rates of the various partial reactions of photosynthesis can have significant effects on the expected stromal concentrations of these transported intermediates. However, the relative rates of the partial reactions are actually not independent of the external metabolite levels, as will be shown below.

Effects of External Metabolites on Starch Synthesis. The marked changes in the activity of ADP-glucose pyrophosphorylase by Pi and PGA, observed in experiments with the isolated enzyme (19), suggested that changes in these cell metabolites might be a primary factor in the regulation of starch synthesis. The ability of external Pi to alter the rate of starch synthesis in isolated chloroplasts has been reported by several groups (11, 20). Low external Pi favored starch synthesis and the percent of total fixation diverted to starch correlated very well with measurements of the stromal PGA:Pi ratio.

However, manipulation of only the external Pi concentration caused marked changes in both starch synthesis and total fixation, and the highest rates of starch synthesis were observed at Pi concentrations which are limiting for photosynthesis. While high rates of starch synthesis from either external PGA or DHAP could be observed (11), the relationship between starch synthesis, external metabolites, and total fixation was unclear in several respects.

We have investigated the effects of relatively high and more physiological concentrations of Pi, PGA, and DHAP on starch synthesis with isolated chloroplasts (17). As shown in Figure 2, with 3 mM PGA present in the external medium, the external Pi could be varied from 1 to 6 mM with a rapid reduction in the rate of starch synthesis, but with little change in total fixation. Similar results (not shown, but see Ref. 16) were obtained with mM concentrations of DHAP or of both PGA and DHAP, which would be the closest to the in situ situation.

The regulation of starch synthesis independently of CO_2 fixation is also clearly demonstrated by similar experiments conducted at low-light intensity. In contrast to experiments with Pi alone (11), starch synthesis is not very sensitive to light intensity when both PGA and Pi are present in the external medium (Table I). The high rate is, however, supported by external PGA and not newly fixed CO_2, but the actual proportions cannot readily be determined because of isotopic dilution effects in this type of experiment. In contrast, when DHAP and Pi are present, starch synthesis is more dependent on light than is CO_2 fixation. This result is probably explained by the fact that under low light, the rate of formation of PGA in the stroma is much reduced and with very low levels of PGA existing in the medium, the stromal PGA concentration is certainly much less than under high-light conditions.

In these experiments, we have only looked at new starch synthesis and not net starch production. The effects of external metabolite ratios (16, 20-22), on starch degradation have not been as extensively investigated. Nevertheless, it is clear that starch synthesis is only casually related to photosynthesis. Observations of net starch synthesis in the dark (13) and breakdown in the light (22) may be eventually explained by the appropriate stromal and cytoplasmic metabolite levels existing under these conditions.

Effect of External Metabolites on the Absolute Rate of Photosynthesis. Isolated chloroplast photosynthesis is very sensitive to external Pi. However, as shown in principle in very early studies (3), other

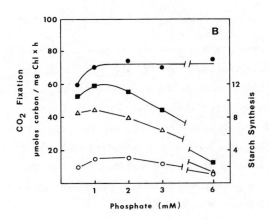

FIG. 2. Effect of external phosphate and 3-phosphoglyceric acid on CO_2 fixation and starch synthesis. CO_2 fixation (●) and starch synthesis (■) with 3 mM external PGA were measured. Starch synthesis was also measured by incorporation of label from $[^{14}C]$-PGA (△) or $[^{14}C]$-CO_2 (○) in parallel reaction mixtures, as previously shown (17).

TABLE I. Effect of Light Intensity on Starch Synthesis

	Light Intensity	CO_2 Fixation	Starch Synthesis		
	(μmol m^{-2} s^{-1})		(μmol mg^{-1} Chl h^{-1})		
Experiment #1			[^{14}C]-PGA	$^{14}CO_2$	Total
1 mM PGA + 1 mM Pi	500	110	6.2	4.2	10.4
	90	45	8.4	1.5	9.9
	30	8	4.9	0.15	5.0
0.3 mM Pi	500	100	–	7.1	7.1
	90	54	–	2.3	2.3
Experiment #2			[^{14}C]-DHAP	$^{14}CO_2$	Total
1 mM DHAP + 1 mM Pi	1000	125	2.5	5.4	7.9
	100	32	0.7	0.2	0.9
	30	15	0.5	0	0.5

metabolites can greatly alter the effect of Pi. Figure 3 illustrates that high absolute concentrations of Pi are not necessarily inhibitory for isolated chloroplast photosynthesis. Inhibition of 50% requires Pi:PGA and Pi:DHAP ratios of over 20:1; a ratio of 6:1 only inhibits 20%. However, as high Pi is inhibitory for various reactions outside of the chloroplast, which leads to sucrose synthesis (1, 4, 5, 9, 23) in situ, a differential sensitivity to Pi may ensure that the cytoplasmic Pi readily falls to a level such that chloroplast export matches sucrose synthesis, as the chloroplast consumes Pi.

At the other extreme, we may inquire as to what extent isolated chloroplast photosynthesis is inhibited by excessive PGA:Pi or DHAP:Pi ratios. As shown in Figure 4, considerable inhibition of photosynthesis can be observed at ratios of triose-P to phosphate as high as 20:1 with either PGA or DHAP, but photosynthesis is still occurring at about 35% of maximal. A large proportion of this remaining rate is certainly accounted for by starch synthesis. However, the inhibition

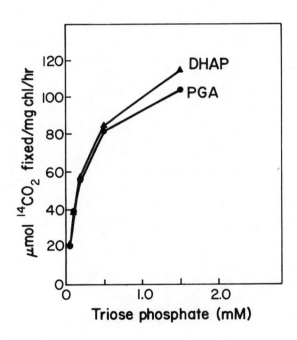

FIG. 3. Reversal of the inhibition of CO_2 fixation by high levels of Pi with dihydroxyacetone phosphate or 3-phosphoglyceric acid. For this experiment, Pi was 10 mM; control rate with 0.3 mM Pi was 133 µmol mg^{-1} Chl h^{-1}.

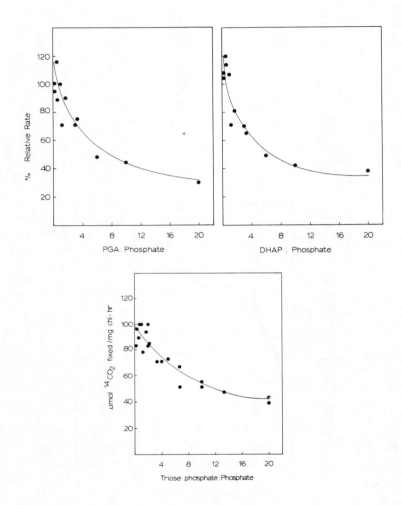

FIG. 4. Influence of external PGA:Pi (top left), DHAP:Pi (top right), or equal (DHAP and PGA):Pi (bottom) ratios on CO_2 fixation. DHAP and PGA varied from 0.5 to 6 mM, and Pi from 0.3 to 12 mM in various reactions. Results are from numerous experiments normalized to controls with 0.3 mM Pi.

is more gradual when both PGA and DHAP are present at equal levels than with either alone. This effect is probably a result of a more favorable redox and metabolite balance under such conditions, but measurements of the stromal metabolites under such conditions are required for a more definitive analysis. Inhibition starts at relatively low triose-P-to-Pi ratios (ca. 3:1). When considered with our data on starch synthesis (Fig. 2; Table I), it would appear that physiological ratios are limited to between 3:1 to 1:3 under normal conditions (i.e. no inhibition of photosynthesis) and may, in fact, be more limited once the implications of changes in these ratios for flux to sucrose become more defined. For a high rate of starch synthesis, the ratio of triose-P:Pi would have to be above 1:1.

Under nonsaturating light conditions, the situation might differ considerably, since the availability of ATP would be expected to be more limiting (10) and, therefore, external metabolites might have a stronger influence on the relative rates of the partial reactions of photosynthesis. We have investigated the effects of low light on optimal PGA:DHAP ratios and triose-P:Pi ratios to some extent. As shown in Figure 5, both PGA reduction and DHAP utilization for CO_2 fixation occur to a greater extent relative to normal CO_2 fixation with an expected $CO_2:O_2$ ratio of 1.0, at high and low PGA:DHAP ratios, respectively. Balance between these extremes appears to require a PGA:DHAP ratio of about 3:1 at intensities adequate for one-half maximal photosynthesis (200 µmol m^{-2} s^{-1}) and becomes much less distinct at high light with much higher or lower ratios being required to observe inhibition (2, 15). Data at even lower light intensities is being pursued, but the experiments require more precision.

In conclusion, we have presented both some theoretical calculations and some experimental data obtained with isolated chloroplasts relevant to several aspects of the relationships between external metabolites, stromal metabolites, the rate of photosynthesis and assimilate partitioning. From such studies, we can begin to define the "expected" physiological range over which the metabolite ratios of Pi:PGA, Pi:DHAP, and PGA:DHAP could vary based on isolated chloroplast metabolism. These results are of particular importance because of current difficulties

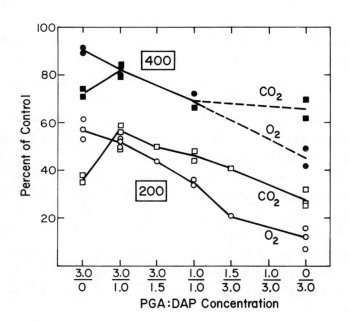

FIG. 5. Effect of external PGA:DHAP concentration ratio on CO_2 fixation and O_2 evolution. CO_2 fixation (■, □) and O_2 evolution (●, o) were measured at light intensities of 400 (■, ●) and 200 (□, o), μmol m^{-2} s^{-1} with 1 mM external Pi. Results are combined from six separate experiments and normalized to controls measured at 1000 μmol m^{-2} s^{-1} and 0.3 mM Pi.

in accurately measuring cytoplasmic Pi levels in protoplast or intact leaf experiments (24).

LITERATURE CITED

1. AMIR J, J PREISS 1982 Kinetic characterization of spinach leaf sucrose phosphate synthase. Plant Physiol 69:1027-1030
2. BAMBERGER ES, BA EHRLICH, M GIBBS 1975 The glyceraldehyde 3-phosphate and glycerate 3-phosphate shuttle and carbon dioxide assimilation in intact spinach chloroplasts. Plant Physiol 55:1023-1030
3. COCKBURN W, DA WALKER, CW BALDRY 1968 Photosynthesis by isolated chloroplasts. Reversal of orthophosphate inhibition by Calvin-cycle intermediates. Biochem J 107:89-95
4. CSEKE C, A BALOGH, BB BUCHANAN, M STITT, HW HELDT 1984 Regulation of the synthesis and breakdown of fructose-2,6-bisphosphate in leaves. C Sybesma, ed. Adv Photosynth Res III, pp 625-628
5. DOEHLERT DC, SC HUBER 1983 Regulation of spinach leaf sucrose phosphate synthase by glucose-6-phosphate, inorganic phosphate, and pH. Plant Physiol 73:989-994
6. FLIEGE R, U FLUGGE, K WERDAN, HW HELDT 1978 Specific transport of inorganic phosphate, 3-phosphoglycerate and triosephosphates across the inner membrane of the envelope in spinach chloroplasts. Biochim Biophys Acta 502:232-247
7. FLUGGE UI, J GERBER, HW HELDT 1983 Regulation of the reconstituted chloroplast phosphate translocator by an H^+ gradient. Biochim Biophys Acta 725:229-237
8. GIERSCH C 1977 A kinetic model for translocators in the chloroplast envelope as an element of computer simulation of the dark reaction of photosynthesis. Z Naturforsch 32c:263-270
9. HARBRON S, C FOYER, D WALKER 1981 The purification and properties of sucrose phosphate synthase from spinach leaves: The involvement of this enzyme and fructose bisphosphatase in the regulation of sucrose biosynthesis. Arch Biochem Biophys 212:237-246
10. HEBER U 1973 Stoichiometry of reduction and phosphorylation during illumination of intact chloroplasts. Biochim Biophys Acta 305:140-152

11. HELDT HW, CJ CHON, D MARONDE, A HEROLD, ZS STANKOVIC, DA WALKER, A KRAMINER, MR KIRK, U HEBER 1977 Role of orthophosphate and other factors in the regulation of starch formation in leaves and isolated chloroplasts. Plant Physiol 59:1146-1155
12. HELDT HW, A GARDEMANN, R GERHARDT, B HERZOG, M STITT, W WIRTZ 1984 The regulation of CO_2 fixation and of sucrose synthesis in plants. C Sybesma, ed. Adv Photosynth Res III, pp 617-624
13. HEROLD A 1978 Starch synthesis from exogenous sugars in tobacco leaf discs. J Exp Bot 29:1391-1401
14. HERZOG B, M STITT, HW HELDT 1984 Control of photosynthetic sucrose synthesis by fructose 2,6-bisphosphate III. Properties of the cytosolic fructose 1,6-bisphosphatase. Plant Physiol 75:561-565
15. KAISER W, W URBACH 1976 Rates and properties of endogenous cyclic photophosphorylation of isolated intact chloroplasts measured by CO_2 fixation in the presence of dihydroxyacetone phosphate. Biochim Biophys Acta 423:91-102
16. PEAVEY DG, M STEUP, M GIBBS 1977 Characterization of starch degradation in intact spinach chloroplast. Plant Physiol 60:305-308
17. PORTIS AR 1982 Effects of the relative extrachloroplastic concentrations of inorganic phosphate, 3-phosphoglycerate and dihydroxyacetone phosphate on the rate of starch synthesis in isolated spinach chloroplasts. Plant Physiol 70:393-396
18. PORTIS AR 1983 Analysis of the role of the phosphate translocator and external metabolites in steady-state chloroplast photosynthesis. Plant Physiol 71:936-943
19. PREISS J 1982 Regulation of the biosynthesis and degradation of starch. Annu Rev Plant Physiol 33:431-454
20. STEUP M, DG PEAVEY, M GIBBS 1976 The regulation of starch metabolism by inorganic phosphate. Biochim Biophys Res Commun 72:1554-1561
21. STITT M, HW HELDT 1981 Physiological rates of starch breakdown in isolated intact spinach chloroplasts. Plant Physiol 68:755-761
22. STITT M, HW HELDT 1981 Simultaneous synthesis and degradation of starch in spinach chloroplasts in the light. Biochim Biophys Acta 638:1-11

23. STITT M, B HERZOG, R GERHARDT, B KURZEL, HW HELDT, C CSEKE, BB BUCHANAN 1984 Regulation of photosynthetic sucrose synthesis by fructose 2,6-bisphosphate. C Sybesma, ed. Adv Photosynth Res III, pp 609-611
24. STITT M, W WIRTZ, HW HELDT 1980 Metabolite levels during induction in the chloroplast and extrachloroplast compartments of spinach protoplasts. Biochim Biophys Acta 593:85-102

Enzyme Assays at the Single-Cell Level: Real-time, Quantitative, and Using Natural Substrate in Solution

WILLIAM H. OUTLAW, JR., STEFAN A. SPRINGER, and MITCHELL C. TARCZYNSKI

A higher organism is made of distinctly different types of cells each of which may be presumed to fulfill particular functions in the organism's metabolism. It is unfortunate that advances in our understanding of biochemistry at other levels of biological organization have not been paralleled at the level of the cell. While there are notable exceptions to this generality (e.g. the specialization of cells in C_4 photosynthesis), it applies particularly to plant sciences. Our experience in guard cell physiological investigations indicates that a major impediment to progress is the technical difficulty of applying single-cell techniques that result in reliable, quantitative data. As the enzyme complement of a cell is one of its most basic attributes, we have devoted considerable effort to the development of assay protocols that should be generally applicable. In this regard, the standards of a cytochemist should be the same as those of any other biochemist: ideally, (a) the assay indication should be in real time, providing a record of the kinetics of the reaction, (b) the assay substrate should be the naturally occurring one, and (c) the components of the assay cocktail should be in solution at desired concentrations. In practice, these constraints require that the method for estimation of the reaction progress should be nondestructive and, to achieve a time resolution of seconds, should have a sensitivity of femtomoles.

Our recent paper (2) reported a method that satisfies all the above criteria; that paper should be consulted for references to earlier work. The primary report (2) was a detailed, technical one, directed to specialists in quantitative histochemical analysis. Because of space constraints, the methods for tissue manipulation—which were developed by Lowry and co-workers (1)—were not included. Thus, the purpose of this paper is to provide an overview of the entire set of methods for the general reader. To do this, we omit distracting minutiae and cautionary notes required to implement the procedure. First, we briefly describe methods for tissue sampling, and then mass determination of single cells. Next, an actual procedure for assaying phosphoenolpyruvate carboxylase in a single guard cell pair of <u>Vicia faba</u> is presented. Last, we discuss data interpretation. An appendix that provides answers to frequently asked questions follows the manuscript.

PREPARATION OF TISSUE SAMPLES

Tissue from which samples will be derived are subjected to a physiological treatment (<u>e.g</u>. light), if desired. Then, the tissue is rapidly quenched in N_2 that has been cooled to its melting point by evacuation. At this point, the procedure differs, depending on the nature of the tissue. Tissues that are thin and relatively friable when dry are crumbled in CO_2 powder and directly freeze-dried. Solid or compact tissues are sectioned first on a cryostat, and then freeze-dried. Freeze-drying not only stabilizes enzyme activities within the tissue, but it also disorganizes membranes so that they are no longer barriers to diffusion of low mol wt substances.

Manual dissection of small freeze-dried samples is more easily accomplished than one might expect. A razor blade fragment is mounted on a nylon bristle, which, in turn, is cemented to a copper wire on the end of a dowel handle. (The wire allows one to position the blade while the bristle dampens hand movement.) Using this inelegant tool, most people are able to dissect with a morphological resolution of 1 to 2 µm. Samples are also transferred manually, with a small quartz fiber attached to a handle. Figure 1 is a photograph of individual palisade parenchyma cells, and individual spongy parenchyma cells, which were arranged for effect under an ordinary stereomicroscope.

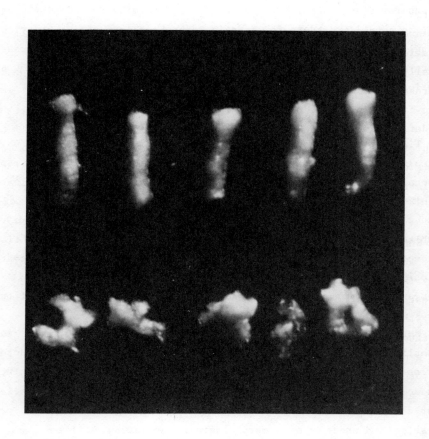

FIG. 1. Palisade parenchyma cells (top) and spongy parenchyma cells (bottom) dissected from a frozen-dried *Vicia faba* leaflet. The cells are arranged regularly to demonstrate the ease with which small tissue samples can be manipulated by hand. (Reprinted by permission of the American Society of Plant Physiologists.)

DETERMINATION OF CELL MASS

Presently, there are two primary choices for expression of specific activity: mass or cell basis. The quartz fiber "fishpole" balance, which may be constructed to have a sensitivity of picograms, is depicted in Figure 2. This device is not ideal because it has a restricted range over which sufficiently precise measurements can be made. However, its positive attributes offset this inconvenience: it is durable and inexpensive and requires virtually no maintenance. Furthermore, the design and operation are straightforward. A very thin quartz fiber is mounted horizontally in a small case where static electricity and wind are eliminated. The resting position of the fiber is determined with a stereomicroscope. After tissue is transferred to the fiber tip, its position is determined. The deflection of the fiber due to the weight of the sample is used to calculate the sample's mass. Despite the simplicity of the balance, its precision and accuracy are within a few percent; this source of error is the largest in an assay.

ASSAY PROCEDURE

General.--An assay is conducted in a microdroplet of reaction cocktail, which is suspended under oil to prevent evaporation (Fig. 3). The assay indicator is the specific oxidation or reduction of NAD(P) (4). As these compounds are utilized by or can be coupled to most enzyme activities, there are few constraints on the design of assays. The reaction is continually monitored by sequential measurements of NAD(P)H fluorescence with a microscope photometer and computer.

Demonstration assay: Phosphoenolpyruvate carboxylase in a single guard cell pair of Vicia faba.[1] The endogenous activity of this enzyme (A) is coupled with analytical malic dehydrogenase (B) in excess. With some components omitted for simplicity, the reaction may be written as in eqn. 1.

---

[1] Perhaps it should be noted that, on an absolute [or cell] basis, phosphoenolpyruvate carboxylase in a guard cell pair is not unusually high compared to "typical" enzyme activities in single cells.

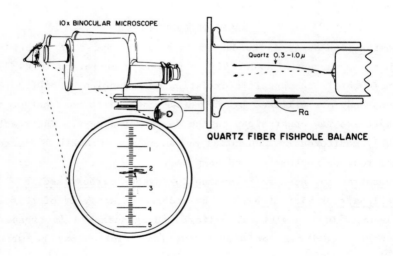

FIG. 2. Quartz fiber "fishpole" balance. The deflection toward gravity of a horizontally-mounted fiber is proportional to the mass of tissue resting on the fiber tip. See text for further details. (Courtesy of Prof. Oliver H. Lowry.)

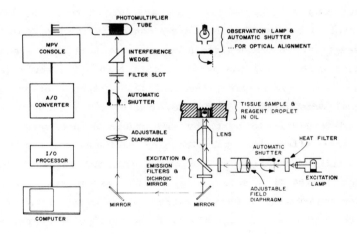

FIG. 3. Diagram of instrumentation used to measure fluorescence of femtomole quantities of NAD(P)H in nanoliter droplets of enzyme reaction cocktail. (Reprinted by permission of the American Society of Plant Physiologists.)

$$\text{phosphoenolpyruvate} + CO_2 + H_2O \xrightarrow{(A)} \text{oxaloacetate} + Pi \xrightarrow[NADH \quad NAD^+]{(B)} \text{malate} \quad [1]$$

Since NADH (but not NAD^+) fluoresces, cocktail fluorescence loss is a quantitative indicator of oxaloacetate formation.

Before actually conducting the assay itself, the investigator sequentially completes some preliminary procedures, as demonstrated below, by viewing the computer display screens during the progression of MICROENZYME (our software for data acquisition, processing, and storage). The first display provides for an input of general assay parameters (Fig. 4). The first three lines are simply for recordkeeping, while inputs for lines four (cocktail volume) and five (concentration of NAD(P)H for calibration) are used for subsequent calculations. From the input on line six (precision), an empirically-derived equation resident in the software is used to calculate the number of amplifier updates (or individual digitization) of the photometer signal required for the desired precision. In the present example, 89 updates were averaged into each datum. This input is critical because it will place a constraint on the time resolution of the assay (e.g. if the desired precision required 5000 signal updates averaged into each datum, the possible time resolution would deteriorate to about 125 s). Furthermore, the excitation light shutter opens only during signal acquisition. Therefore, the extent of photobleaching of NAD(P)H is determined, in part, by the precision one sets. The other parameters that affect the photobleaching estimate are set in the last two lines (maximum duration of the assay and time resolution). Before the operator leaves this screen, any of the parameters can be changed freely. For convenience, the operator is prevented from leaving the screen if the inputs are illogical (e.g. the appropriate warning is given if the time resolution exceeds the assay duration) or inconsistent with hardware capabilities (e.g. if the time required to access and process extremely high precision data exceeds the desired frequency, a warning and an estimate of the maximum frequency possible at that precision will be written over the operator's original input). In practice, setting of these parameters and their adjustment, if necessary, can be completed in

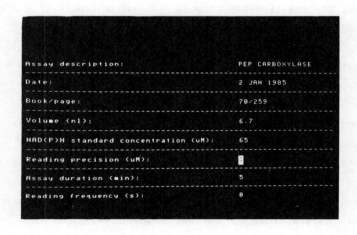

FIG. 4. First screen of MICROENZYME, a computer program for acquisition, processing, and storage of data (see FIG. 3).

seconds. For the example given here, after the screen was set as shown (Fig. 4), the procedure was continued.

The next screen (Fig. 5) provides an estimate of photobleaching during the assay. If that estimate is unacceptably high, the operator must loop back to the first screen (Fig. 4) and alter either time resolution or measurement precision. Otherwise, the next screen of MICROENZYME appears (Fig. 6), which provides for calibration. In the current example, the photobleaching estimate was accepted, but this initial estimate later was shown on the printout to be too high (the data were calculated over an interval shorter than the set duration, and NADH concentration was declining because of oxaloacetate reduction).

For this demonstration, ca. 7 nl of assay cocktail with NADH omitted was suspended in the oil of the assay chamber using a tiny custom-fabricated constriction pipette (left panel of Fig. 7). The microdroplet was then aligned with the inverted lens of the microscope photometer (Fig. 3). On command, excitation light, provided by epi-illumination, passed through the thin Teflon film of the chamber and into the microdroplet. The "blank" fluorescence signal thus acquired was incorporated into memory. The procedure was repeated for cocktail containing 65 μM NADH. Thus, the relationship between NADH concentration and fluorescence was incorporated into memory.

After calibration, a new screen appears, which provides for input of specific data regarding the particular cell sample (Fig. 8). Given logical inputs, the screen is replaced by a plotting frame (not shown, similar to Fig. 9) indicating that the assay itself can begin. (Instructions executed by soft keys, such as "abort" or "proceed", are presented to the right of the plotting frame.)

We initiated this assay by transferring a weighed guard cell pair to a droplet of complete assay cocktail (right panel of Fig. 7). (Needless to say, the same control procedures required in "macro" assays pertain, but to avoid distraction, no further comments will be made.) With the reagent aligned on the microscope photometer, signal acquisition began, and the reaction progress was updated on the screen every 8 s. The reaction could have been aborted, in which case MICROENZYME would have accepted data for a new cell (Fig. 8). It also is possible to terminate

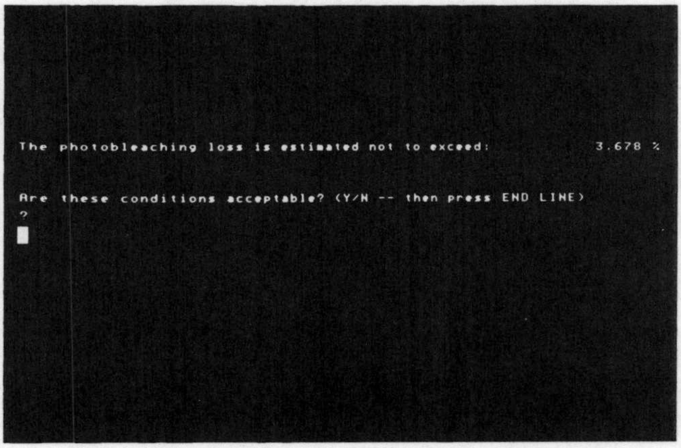

FIG. 5. Second screen of MICROENZYME.

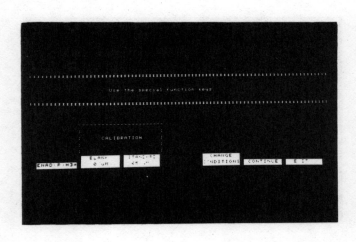

FIG. 6. Third screen of MICROENZYME.

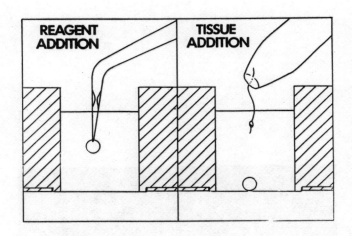

FIG. 7. Cross section of the reaction vessel used for assay of enzyme activities in single cells. Holes drilled through a solid Teflon block are sealed on the bottom with a thin Teflon film. The resulting wells are filled with oil. Nanoliter quantities of assay cocktail are delivered into the oil by use of small custom-fabricated constriction pipettes (left panel). The reaction is initiated when tissue is introduced into the assay cocktail (right panel). The reaction progress is followed by sequential measurements of NAD(P)H fluorescence using an inverted microscope photometer. See Figure 3 and text.

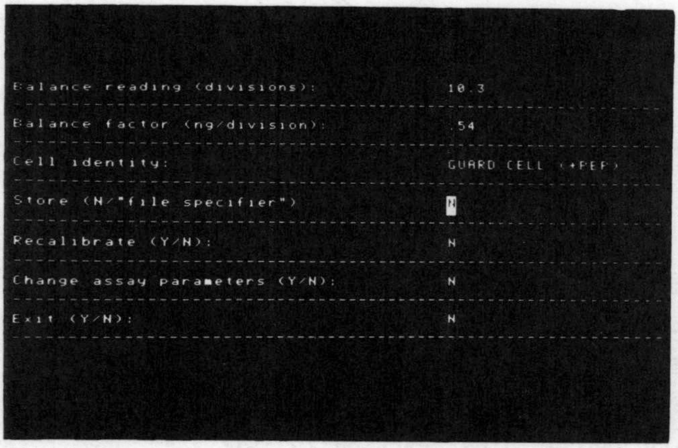

FIG. 8. Fourth screen of MICROENZYME.

a reaction before the set duration expires, in which case MICROENZYME would continue as indicated for the "normal" case below.

At the end of an assay, the previous screen is modified so that the operator may select limits for calculations (Fig. 9). In this example, the reaction seemed to lose linearity after about 3.5 min. (The loss of linearity was due to NADH depletion; the baseline of the plotting frame is set between dark current of the instrument and the fluorescence signal of cocktail with NADH omitted.) Thus, the calculation brackets (near top of Fig. 9) were positioned by soft keys to exclude the last 1.5 min of the assay. For the sake of demonstration, the left bracket was displaced from the origin. At this point, the assay was completed. After the regression was drawn (not shown) for the "valid" points, a hard copy of the graphics and other information was produced on the system printer. Data storage, an option passed over (Fig. 8), would have been a possibility. The computer returns to accept data for a new cell (Fig. 8).

Data Interpretion.--A successful investigation is complete only after it appears in the literature. In some cases--like the one demonstrated here--the results are straightforward. It should be enough to conduct the assays, perhaps show a typical one, and report the mean specific activity. For a variety of reasons, however, the quality of the data may deteriorate. As we have indicated, if very high time resolution is desired, reading precision must be sacrificed. Other reasons for loss of data quality would include low endogenous activity (either at V_{max} or during kinetic characterization), a desire to lower photobleaching of NAD(P)H, and measurement at low NAD(P)H concentration (e.g. to show constancy of activity at high tissue dilution). Thus, our software repertoire includes a program, which we dub CONSOLIDATE, that can combine up to 100 assays of the kind shown here. (Each assay is limited to 10,000 X-Y data pairs, each of which may be the average of up to 5,000 amplifier signals). In the present context, the important capabilities of CONSOLIDATE are (a) its independence of the time resolution of different experiments, (b) the ability to collapse data into sets at lower time resolution, (c) the ability to weight experiments differently, (d) the ability to correct for baseline shifts, and (e) the ability to correct for amplification shifts. While one hopes that data manipulation is unnecessary, it may extend the

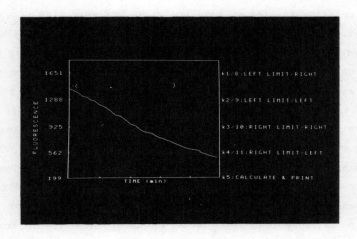

FIG. 9. Fifth screen of MICROENZYME.

sensitivity considerably. We believe that this capability means that virtually any enzyme that can be coupled to NAD(P) can be measured at the levels likely to be significant in a cell's metabolism. The final program in this set, ARRAYPLOT, retrieves data stored by CONSOLIDATE or MICRO-ENZYME and produces a publication-quality plot (see examples in Ref. 2).

CONCLUSIONS

Using phosphoenolpyruvate carboxylase in a guard cell pair as an example, we have shown that enzyme activities in single cells can be measured in real time at a resolution of seconds using natural substrates in solution. We believe that these methods are a significant improvement over other methods for quantifying minute quantities of enzyme activity. In addition, the approach opens up some new avenues for investigating cell function (e.g. enzyme characterization at the single-cell level).

ACKNOWLEDGMENT--Supported by a grant from the U.S. Department of Energy.

LITERATURE CITED

1. LOWRY OH, JV PASSONNEAU 1972 A flexible system of enzymatic analysis. Academic Press, New York.
2. OUTLAW WH JR, SA SPRINGER, MC TARCYNSKI 1985 Histochemical technique: a general method for quantitative enzyme assays of single-cell "extracts" with a time resolution of seconds and a reading precision of fmoles. Plant Physiol 77:648-652

APPENDIX

<u>Can Enzyme X be measured in individual Y-type cells?</u> The two parts of this question must be answered separately. If the enzyme can be assayed spectrophotometrically with a NAD(P) link in raw extracts, it is likely that the assay can be miniaturized. If the minimum dimension of a Y-type cell is 10 μm or greater, it is likely that it can be sampled with sufficient precision.

<u>How do these assays differ from a normal spectrophotometric assay?</u> The most important difference is that there is no means of freeing single-cell "extracts" of endogenous low-mol-wt substances; instead, in the microassay, one dilutes the extract more.

<u>Can metabolites also be measured using these principles?</u> Yes, but there is no advantage over enzymatic cycling techniques (1).

<u>How difficult are the techniques to learn?</u> That varies with individuals. Despite the technical nature of the approach, the most important requirement is a general knowledge of enzyme analysis. Beyond that, working exposure to an experienced individual is important, if not essential. We feel that, after three months, an individual will have a sufficient basis to determine whether he/she should continue. In other words, it would be inappropriate to list a to-be-announced postdoc in a grant application as the one having primary responsibility for conducting a project of this nature.

<u>How long does it take to do an analysis?</u> As we have indicated, a single assay for which one is set up can be conducted fairly rapidly. The bulk of the time, say six weeks, is involved in optimizing the assay, reducing contaminants in reagents, and validating the analysis.

<u>How much does it cost to set up for these assays?</u> About $70,000 for the microscope fluorometer, computer, and construction of the dissecting room. That figure, of course, does not include equipment that is likely to be present already.

How long does it take to set up a laboratory? A reasonable estimate is 12 months. There is a fair amount of custom fabrication, including some the investigator must do himself (e.g. "fishpole" balances). Unless the investigator is fairly diverse, he will also need help from mechanical, hardware, and software specialists.

Carboxylase Response to CO_2 and O_2 in Intact Leaves of Wheat and Maize

STEVEN W. GUSTAFSON, DEBORAH A. RAYNES, and RICHARD G. JENSEN

Little is known of the response of ribulose 1,5-bisphosphate (RuBP) carboxylase/oxygenase (Rubisco) in leaves exposed to a CO_2-free atmosphere. The ability of O_2 to support activation of Rubisco in the absence of CO_2 was reported by Perchorowicz and Jensen (1). They suggested the O_2 effect on activation was facilitated by turnover of RuBP in photorespiration and its regeneration in the Calvin cycle, thus using $NADPH_2$ and maintaining electron flow through the photosystems. However, the role of photorespiration in maintaining activation was questioned since changes in activation occurred at O_2 levels thought too low to support photorespiration (1, 4).

Rubisco activation rates of wheat seedlings exposed to an N_2 atmosphere were found to be near zero while the RuBP concentration was at "binding-site" levels (1). This was interpreted to indicate that the inactive enzyme was mostly bound to RuBP, thus maintaining RuBP concentrations at "binding-site" levels or above.

Exposure to N_2 in the absence of CO_2 and O_2 often causes photoinhibition (4) and could have resulted in the incomplete recovery of activation for the above-mentioned wheat leaves when CO_2 and O_2 were restored (1). Photorespiration has been hypothesized to prevent or minimize photoinhibition in tissue having low CO_2 by maintaining a minimal CO_2 level in order to provide activation of Rubisco.

In order to better understand the relationship between O_2, RuBP levels, photorespiration, and Rubisco activation and their possible roles in regulating photoinhibition, experiments were conducted to measure the effect of exposing wheat and maize seedlings to either 1% O_2, or only N_2 for a 3-h period followed by 350 ppm CO_2 and 21% O_2 on net photosynthesis, RuBP levels and Rubisco activity.

MATERIALS AND METHODS

Durhum wheat var., Mexicali and sweet corn var. Golden Cross Bantam T51 were grown from seed in 1:1 (v/v) potting soil/vermiculite in 397 ml styrofoam pots within a growth chamber (environmental conditions: 16-h photoperiod, 30° day/20° night, PAR = 400 µmol photons m^{-2} s^{-1} at pot level, and 50% RH). Plants were used after 7 to 8 d of growth. Gas exchange measurements were made using a system modified from that described by Perchorowicz et al. (2). Gases were mixed manually, using multiple valves and flow mixtures. Changes in mixtures were accomplished within 4.0 min. Light intensities averaged 1300 µmol photons m^{-2} s^{-1}. Rubisco and RuBP levels were determined as previously described (1, 3).

RESULTS

Wheat seedlings exposed to 1% O_2 for 3 h did not exhibit photoinhibition. Net photosynthesis recovered to pretreatment levels of 100 µmol CO_2 mg^{-1} Chl h^{-1}. However, photoinhibition was observed in wheat seedlings exposed to N_2 alone, and caused a 60% reduction in net photosynthesis. Maize seedlings in N_2 suffered only 20% photoinhibition as net photosynthesis reached 80 µmol CO_2 mg^{-1} Chl h^{-1} in 1.5 h after air levels of CO_2 and O_2 were restored.

RuBP levels increased during all treatments. At the end of the 3-h treatment with wheat, the levels were 2 to 3 times greater than the pretreatment levels (120 nmoles mg^{-1} Chl). Even greater increases in RuBP levels occurred in maize leaves under both treatments. When air levels of CO_2 and O_2 were restored, RuBP in leaves of both maize treatments and the 1% O_2-treated wheat returned to pretreatment levels. However, RuBP levels in N_2-treated wheat leaves declined below pretreatment (presumed to be "binding-site") levels.

The initial activity of Rubisco decreased in both species under both treatments. However, initial activity in N_2-treated wheat approached zero within 1.5 h after the treatment began. In contrast, initial activity in

1% O_2-treated wheat and maize under either treatment decreased to only 50%. Upon reexposure to air levels of CO_2 and O_2, the initial activity of both treatments in wheat and maize returned to or above the pretreatment level. This indicated that the inhibition of photosynthesis was not due to inactive Rubisco.

In wheat, total activity levels underwent transient fluctuations but changed little throughout the experiments. However, total activity of the 1% O_2-treated maize leaves nearly doubled. In N_2-treated maize, total activity was 1.7 times greater than the pretreatment level (1.6 μmol CO_2 mg^{-1} Chl min^{-1}). Apparently these treatments in maize are able to cause activation of previously inactivated carboxylase. It is doubtful that this represents significant new synthesis of Rubisco proteins within the 3 h.

DISCUSSION

In a CO_2-free atmosphere, 1% O_2 maintained Rubisco activation. In the absence of O_2, Rubisco activity decreased dramatically, especially in wheat. When wheat plants treated in an N_2-atmosphere were exposed to air levels of CO_2 and O_2, the observed decrease in net photosynthesis was due to photoinhibition during the treatment period, as evidenced by the decrease and ability to regenerate RuBP. That O_2 was required to maintain Rubisco activity is compatible with the hypothesis that photorespiration might be preventing photoinhibition. However, this hypothesis has several criticisms, including: does the effect seen at 1% O_2 reflect photorespiration, dark respiration and/or other processes; is the energy-dissipating potential of photorespiration sufficient to prevent photoinhibition; and do plants experience CO_2 levels below their compensation point? Whatever the answer, our results demonstrate that maintenance of some minimum carbon metabolism is required to prevent photoinhibition (5). Maintenance of Rubisco would be essential to such metabolism whether it be through the Calvin-Benson Cycle or the photorespiratory pathway. The mechanisms by which minimal carbon metabolism prevents photoinhibition and the extent of Rubisco activity required is unclear and needs further investigation.

LITERATURE CITED

1. PERCHOROWICZ JT, RG JENSEN 1983 Photosynthesis and activation of ribulose bisphosphate carboxylase in wheat seedlings. Plant Physiol 71:955-960
2. PERCHOROWICZ JT, DA RAYNES, RG JENSEN 1981 Light limitation of photosynthesis and activation of ribulose bisphosphate carboxylase in wheat seedlings. Proc Natl Acad Sci USA 78:2985-2989
3. PERCHOROWICZ JT, DA RAYNES, RG JENSEN 1982 Measurement and preservation of the in vivo activation of ribulose 1,5-bisphosphate carboxylase in leaf extracts. Plant Physiol 69:1165-1168
4. POWLES SB 1984 Photoinhibition of photosynthesis induced by visible light. Annu Rev Plant Physiol 35:15-44
5. POWLES SB, G CORNIC, G LOUASON 1984 Photoinhibition of in vivo photosynthesis induced by strong light in the absence of CO_2: an appraisal of the hypothesis that photorespiration protects against photoinhibition. Physiol Veg 22:437-446

Chloroplasts at Air Levels of CO_2: Factors Influencing CO_2-fixation

RICHARD E. B. SEFTOR and RICHARD G. JENSEN

Research on CO_2 fixation using isolated chloroplasts is usually conducted at saturating levels of CO_2 in order to achieve the maximum rate of CO_2 fixation. This high rate of fixation, however, declines greatly after 10 to 20 min. We utilized an apparatus which permits chloroplasts to photosynthesize under CO_2 concentrations equivalent to air (10 µM aqueous equivalent) without depletion of the CO_2 in order to better understand photosynthesis and the reasons for the decline of CO_2 fixation. In particular, the relationship between photosynthetic CO_2 fixation under air levels of CO_2 and the activities of the ribulose 1,5-bisphosphate carboxylase/oxygenase (Rubisco) has been investigated.

MATERIALS AND METHODS

Chloroplast Preparation and Measurement of The Initial and Total Activities of Rubisco. Chloroplasts were prepared from 6 to 8 week-old spinach plants (Spinacia oleracea var. Viroflay), and the initial and total activities determined as before (4). Determination of the Ribulose-1-5-bisphosphate (RuBP) concentration was according to Jensen and co-workers (2, 3).

Maintaining Chloroplasts at Air Levels of CO_2 (10 µM). Chloroplasts were maintained at air levels of CO_2 by a modification of the procedure used by Stumpf and Jensen (4). This modification involved shaking the chloroplast suspension in a closed vessel with a separate CO_2 buffer which kept the CO_2 concentration in the atmosphere adjusted so that the CO_2 dissolved in the sample wells remained at ca. 10 µM.

Measurement of the activity of the Rubisco as a function of the enzyme complex. The measurement of the activity of the Rubisco activity in terms of the ECM form of the enzyme (see Fig. 1, ref. 1) was made by injecting a sample of the chloroplasts from the chamber into an assay medium of 25 mM HEPES, 1 mM DTE, 0.6 mM RuBP, 10 mM KHCO$_3$ and 0.3 mM EDTA. The EDTA was in sufficient concentration to bind the free Mg^{+2} brought into the assay mixture with the sample and the RuBP was at a saturating concentration in order to stabilize the ECM form of the Rubisco enzyme and prevent the back reaction of ECM going to EC + M (Fig. 1). The fast reaction of EC + M to ECM should be prevented by the lack of added Mg^{+2}.

RESULTS AND DISCUSSION

Chloroplasts at air levels of CO_2 still ceased fixing CO_2 within 35 to 45 min of illumination even though the concentration of RuBP inside the chloroplasts remained at or above the "binding site" level (see Gustafson et al. this Symposium), and the CO_2 concentration of the sample medium remained at ca. 10 μM (aqueous). The total activity of the Rubisco was measured by breaking the chloroplasts into an assay medium containing 20 mM Mg^{+2} and 10 mM KHCO$_3$ (both saturating) and initiating CO_2 fixation by adding the substrate, RuBP, after an incubation period of about 5 min. This activity measures the E + EC + ECM forms of the enzyme present (Fig. 1) and, under the conditions of these experiments, remained above 300 μmol CO_2 fixed mg^{-1} Chl h^{-1}. The initial activity of the enzyme was measured the same way as the total activity except that no 5-min incubation period preceded the addition of RuBP. This indicated the activity of the enzyme in terms of the EC + ECM forms of Rubisco (see fig. 1). While the initial activity decreased to ca. 30% of its original activity, this activity did not correlate with the cessation of photosynthesis. By measuring the activity of the Rubisco in terms of the ECM form of the enzyme, we found that this activity more closely related to the change in CO_2 fixation observed under these conditions. Furthermore, when photosynthetic CO_2 fixation ceased, there was no ECM activity inside the chloroplasts.

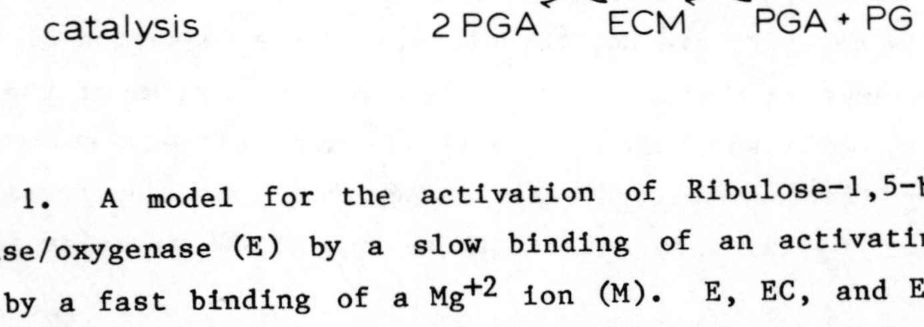

```
activation           inactive  |  active
              slow        fast
         E + C ⇌ EC + M ⇌ ECM
        ─┼─────┼──┼──┼───
  high
   [R]   ER       ECR  C   ECMR    O₂
        ─────────────⤫─────⤫───
catalysis         2 PGA   ECM   PGA + PG
```

FIG. 1. A model for the activation of Ribulose-1,5-bisphosphate carboxylase/oxygenase (E) by a slow binding of an activating CO_2 (C), followed by a fast binding of a Mg^{+2} ion (M). E, EC, and ECM can bind RuBP (R), but only the ECMR species plus CO_2 or O_2 can lead to ^{14}C-labeled products (see Ref. 1 for more details).

Acknowledgments—We would like to thank Deborah Raynes for her expert technical assistance, and both Ms. Raynes and Steven Gustafson for their valuable discussions.

LITERATURE CITED

1. PERCHOROWICZ JT, RG JENSEN 1983 Photosynthesis and activation of ribulose bisphosphate carboxylase in wheat seedlings. Plant Physiol 71:955-960.
2. SICHER RC, JT BAHR, RG JENSEN 1979 Measurement of ribulose 1,5-bisphosphate from spinach chloroplasts. Plant Physiol 64:876-879
3. SICHER RC, RG JENSEN 1979 Photosynthesis and ribulose 1,5-bisphosphate levels in intact chloroplasts. Plant Physiol 64:880-883
4. STUMPF DK, RG JENSEN 1982 Photosynthetic CO_2 fixation at air levels of CO_2 by isolated spinach chloroplasts. Plant Physiol 69:1263-1267

Regulation of Resynthesis of the Photosynthetic CO_2-acceptors: Feedback Inhibition of Transketolase

FRANCIS C. KNOWLES

Upon addition of either crude or purified rabbit skeletal muscle ribose-5-P isomerase (EC 5.3.1.6) to a solution of ribose-5-P in phosphate buffer of neutral pH, an absorption band at 280 nm appears, the rate of production of which decreases continuously until an equilibrium is attained. The 280 nm-absorbing species is the carbonyl moiety of the product of the isomerase catalyzed reaction, ribulose-5-P. Repetition of this experiment with purified spinach chloroplast ribose-5-P isomerase produces a richly-structured spectroscopic display with sequentially-appearing absorption bands at 280, 308.5 and 285 nm. The compound absorbing at 285 nm was shown to be 4-hydroxy-5-methyl-3(2H)-furanone (3) which had previously been isolated from beef broth (5) and prepared from ribose-5-P under mildly acidic conditions (4). The second enzymatic activity, which acted upon ribulose-5-P was assumed to be ribulose-5-P epimerase (EC 5.1.3.1) and the spectroscopic display was attributed to the instability of xylulose-5-P. This assumption was tested by measurement of ribulose-5-P epimerase activity in ribose-5-P isomerase as the isomerase was purified from a specific activity (units/mg^{-1} protein) of less than 0.6 to greater than 2200. Ribulose-5-P epimerase activity was found to disappear from the enzyme preparation, disproving the hypothesis that xylulose-5-P was the precursor of the 285 nm-absorbing species. As ribulose-5-P epimerase disappeared from the ribose-5-P isomerase preparation, the coupled enzyme system used to follow expression of epimerase activity underestimated the production of xylulose-5-P. Failure of the coupled enzyme assay system was traced to an inhibition of transketolase, inhibition being attributed to the precursor of 4-hydroxy-5-methyl-3 (2H)-furanone. In this communication, the inhibition of transketolase is described and the implications of a feedback inhibition mechanism for regulation of resynthesis of CO_2-acceptor are discussed.

METHODS

Two methods were used to measure the activity of ribulose-5-P epimerase. A sampling procedure was carried out in which the quantity of xylulose-5-P formed in a given period of time was determined by spectrophotometric titration, being converted to sedoheptulose-7-P and glycerol-3-P. The quantity of xylulose-5-P present in the sample was calculated from the absorbance excursion at 338 nm due to oxidation of NADH by glycerol-3-P dehydrogenase. A coupled enzyme system was also used in which the rate of oxidation of NADH by glycerol-3-P dehydrogenase was monitored after establishing steady-state concentrations of xylulose-5-P, glyceraldehyde-3-P, and dihydroxyacetone-3-P by the action of ribulose-5-P epimerase, transketolase, and triose-phosphate isomerase.

RESULTS

In preparations of ribose-5-P isomerase where the isomerase/epimerase activity ratio was less than 100, both assay procedures yielded valid indications of epimerase activity. The steady-state assay system was found to underestimate epimerase activity in enzyme preparations enriched in isomerase activity. Cross-plots of epimerase activity determined by the sampling and steady-state procedures demonstrated than an inhibitor of the coupling enzyme mixture was formed in the presence of high relative concentrations of the isomerase. The inhibited coupling enzyme mixture was fully active with glyceraldehyde-3-P. Inhibition of the coupling enzyme mixture was attributed to transketolase. Inhibition of the coupling enzyme mixture was also observed when glyceraldehyde-3-P was converted to glycerate-3-P using glyceraldehyde-3-P dehydrogenase.

CONCLUSIONS

Inhibition of the coupling enzyme mixture is readily attributed to transketolase. Accumulation of the substrate for transketolase, xylulose-

5-P, is demonstrated by the sampling assay procedure. That the inhibition occurs at transketolase, and not later in the sequence of catalyzed reactions, was demonstrated by the full activity of the coupling enxyme mixture of an inhibited reaction when presented with glyceraldehyde-3-P as a substrate. Similar results were obtained when production of glyceraldehyde-3-P by transketolase was detected by oxidation to glycerate-3-P instead of reduction to glycerol-3-P.

This mechanism for regulation of transketolase activity may operate in vivo, as well as in vitro. The anhydrase-like activity of ribose-5-P isomerase was only expressed at high concentrations of ribulose-5-P, suggesting that feedback inhibition of transketolase would restrict resynthesis of CO_2-acceptor only under conditions where the concentration of intermediates of the photosynthesis cycle are sufficiently high that high rates of CO_2-fixation are obtained.

The further metabolic transformations of the inhibitory species, indicated as 3,4-anhydro-ribulose-5-P (3), is not known in detail. The inhibitor has been demonstrated to be a preferred substrate for ribulose-5-P kinase, but the reaction of the kinase product with ribulose bisphosphate carboxylase/oxygenase is not known. Bränden et al. (1,2) have kinetically resolved the carboxylase and oxygenase reactions, demonstrating that the substrate for the oxygenase reaction has a molybdate-enhanced optical rotation opposite in sign to that of D-glycerate-3-P. Thus, the compound indicated to be 3,4-ribulose-1,5-bisphosphate may be an alternate substrate for the photorespiration pathway.

LITERATURE CITED

1. BRÄNDÉN R, T NILSSON, S STYRING 1980 The formation of L-3-phosphoglyceric acid by Ribulose-1,5-bisphosphate carboxylase. Biochem Biophys Res Commun 92:1297-1305
2. BRÄNDÉN R, T NILSSON, S STYRING, J ÅNGSTRÖM 1980 L-3-Phosphoglyceric acid, formed by Ribulose-1,5-bisphosphate carboxylase, is the primary substrate for photorespiration. Biochem Biophys Res Commun 92:1306-1312

3. KNOWLES FC, JD CHANLEY, NG PON 1980 Spectral changes arising from the action of spinach chloroplast ribosephosphate isomerase on ribose-5-phosphate. Arch Biochem Biophys 202:106-115
4. PEER HG, GAM VAN DEN OUWELAND 1968 Synthesis of 4-hydroxy-5-methyl-2,3,dihydrofuran-3-one to D-ribose-5-phosphate. Recl Trav Chim Pays-Bas 87:1017-1020
5. TONSBEEK CHT, AJ PLANCKEN, T WEERDHOF 1968 Components contributing to beef flavor: Isolation of 4-hydroxy-5-methyl-3(2H)-furanone and its 2,5 dimethyl homolog from beef broth. J Agric Food Chem 16:1016-1021

CO_2 Photoassimilation of Heat-Stressed Spinach Chloroplasts

CHEE F. FU and MARTIN GIBBS

The effect of temperature on whole plant photosynthesis has received considerable attention. Clearly, investigation of temperature regulation of photosynthesis requires analysis at all levels of organization that may influence the assimilation of CO_2. In addition to the light-driven fixation of CO_2, temperature may affect photosynthetically-related processes such as stomatal closure and mitochondrial activities, resulting in equivocal conclusions. In the present study, the isolated, intact spinach chloroplast was used as a model system to determine the effect of temperature (from $-15°C$ to $+40°C$) on CO_2 photoassimilation per se. The chloroplast system had the decided advantage that photosynthesis can be measured uncomplicated by respiration and other physiological phenomena of whole organisms. The results of our investigations with the spinach chloroplast demonstrate that photosynthetic response to temperatures lower than the optimum of 25°C is reversible. To the contrary, temperature higher than 25°C eventually results in an irreversible inactivation of a thylakoidal reaction.

MATERIALS AND METHODS

Spinach chloroplasts were isolated as before (2) and were suspended in a 1-ml reaction mixture at ca. 20 µg Chl containing 50 mM Tricine-NaOH (pH 8.1), 0.25 mM KH_2PO_4, 1000 units catalase, 10 mM $NaH^{14}CO_3$ (5 µCi/ml), 2 mM Na_2EDTA, 1 mM $MgCl_2$, 1 mM $MnCl_2$, and 0.33 M sorbitol. The tubes containing this mixture were immersed into a temperature-regulated bath prior to the initiation of illumination (saturating light intensity, 600 w/m^2).

Fixation rates were measured by spotting 10 µl aliquots at appropriate time intervals on planchets containing 0.1 ml of 0.5 M HCl. Radioactivity was determined with a Nuclear Chicago Q-gas end-window counter.

RESULTS AND DISCUSSION

CO_2 **Assimilation at Low Temperature.** Of the intervals (5°C) tested between 0°C and 25°C (optimum temperature), the greatest effect was observed between 0°C and 5°C (Fig. 1). Thus, temperature coefficient (Q_{10}) showed a sharp decrease from 4.4 at 0°C to 5°C to ca. 2 at 20°C to 25°C. A similar trend in Q_{10} has also been observed for a reconstituted spinach chloroplast preparation (not shown).

A characteristic response of our preparations is a linear time course of fixation up to 1 h at the lowest temperatures. The lag time prior to linear rate of fixation is lengthened as the temperature decreases, reflecting the slower buildup in intermediates of the photosynthetic carbon fixation cycle. Finally, CO_2 fixation at a measurable rate of 0.1 µmol mg^{-1} Chl h^{-1} was achieved at -15°C by including 25% ethylene glycol in the incubation medium (Table I).

The CO_2 fixation rate observed at low temperatures (0°C) is resumed to the control rate (25°C) without a noticeable lag upon return to the control temperature (Fig. 2). Inasmuch as the photosynthetic response is reversible, it is clear that the physical components of the chloroplast are not damaged by exposure to low temperature.

CO_2 **Assimilation at High Temperature.** The Q_{10} values for photosynthesis fell off sharply at temperatures higher than the control temperature of 25°C. When CO_2 fixation was performed at 30°C to 35°C, the steady linear rate was sustained for only 10 min (Fig. 3A). At 40°C, fixation ceased after 6 min (Fig. 3C). Transferring of spinach chloroplasts, exposed to 35°C or 40°C for 9 min, to 25°C did not restore fixation (Fig. 3A, 3C). A shorter exposure of 3 min at 35°C (Fig. 3B), but not at 40°C (Fig. 3D), seemed to be reversible.

CONCLUSIONS

The photosynthetic rate of intact chloroplasts was restored fully to the control level even after exposure of the organelles to -15°C or 40

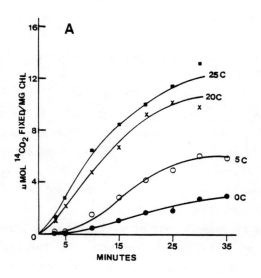

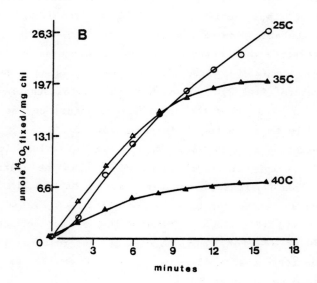

FIG. 1. Temperature effects on $^{14}CO_2$ photoassimilation by spinach chloroplasts. A, $^{14}CO_2$ photoassimilation at low temperature; B, $^{14}CO_2$ photoassimilation at high temperature.

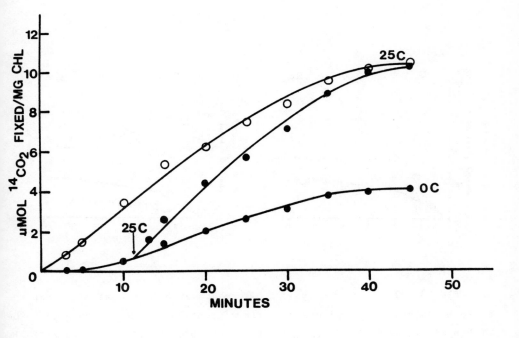

FIG. 2. Reversibility of $^{14}CO_2$ photoassimilation at low temperature by spinach chloroplasts. (See Text for details.)

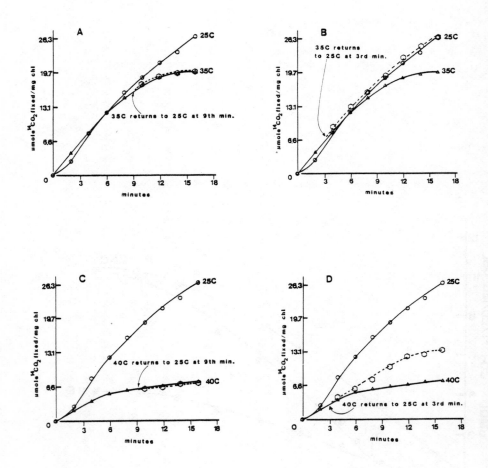

FIG. 3. Irreversibility of $^{14}CO_2$ photoassimilation at high temperature by spinach chloroplasts. A, transfer of 35°C reaction tube to 25°C at 9th min of CO_2 fixation; B, transfer of 35°C reaction tube to 25°C at 3rd min of CO_2 fixation; C, transfer of 40°C reaction tube to 25°C at 9th min of CO_2 fixation; D, transfer of 40°C reaction tube to 25°C at 3rd min of CO_2 fixation. Open circles, 25°C controls.

TABLE I. $^{14}CO_2$ Photoassimilation by Spinach Chloroplasts at −15°C and Response on Return to 25°C

Reaction Temperature	Condition of Illumination	$^{14}CO_2$ Fixation Rate[a] (μmol mg^{-1} Chl h^{-1})
25°C	No add	99.0
25°C	+ ethylene glycol (25%)	7.1
−15°C	+ ethlene glycol	0.1
25°C	+ ethylene glycol @ −15°C for 13'	8.1
25°C	+ ethylene glycol @ −15°C for 46'	6.9

[a]The data are corrected for the dark control values (essentially nil).

degrees below the optimal temperature. On the other hand, elevation of the temperature by as little as 10°C beyond the optimal temperature resulted in an irreversible inactivation and the rate did not return to the control rate of 25°C.

Preliminary experiments with reconstituted spinach preparations established that a thylakoidal, rather than a stromal, component (3) is responsible for heat stress. Inasmuch as the photosynthetic electron transport system remained intact (monitored with the intact chloroplast as O_2 evolution coupled to nitrite or oxaloacetate as electron acceptors), the ATP-coupling factor, among other sites, may well be the focus of heat inactivation.

ACKNOWLEDGMENTS--Supported by National Science Foundation Grant PCM--81-04497.

LITERATURE CITED

1. BALDRY CW, C BUCKE, DA WALKER 1966 Some effects of temperature on carbon dioxide fixation by illuminated chloroplasts. Biochim Biophys Acta 126:207-213
2. BERKOWITZ GA, M GIBBS 1982 Effect of osmotic stress on photosynthesis studied with the isolated spinach chloroplast. Generation and use of reducing power. Plant Physiol 70:1143-1148
3. EMERSON R 1929 Photosynthesis as a function of light intensity and of temperature with different concentrations of chlorophyll. J Gen Physiol 12:623-631
4. VAN DER PAAUW F 1934 Der Einfluß der Temperatur auf Atmung und Kohlensäureassimilation einiger Grünalgen. Planta 22:396-400

Regulation of Sucrose Formation and Movement

STEVEN C. HUBER, PHILLIP S. KERR, and WILLY KALT-TORRES

Plant growth is ultimately dependent on photosynthesis as the source of reduced carbon. Processes that intervene between CO_2 assimilation in mature leaves and the growth of heterotropic plant parts include: (a) carbon partitioning between nontransport (e.g. starch) and transport forms (e.g. sucrose); (b) compartmentation of transport assimilates among pools; (c) phloem loading; (d) long distance translocation; and (e) phloem unloading and uptake/utilization in sink tissues. A comprehensive understanding of plant growth will require basic knowledge of these processes. Our objective is to identify biochemical mechanisms that may regulate sucrose formation and export, and to evaluate how metabolism in "source" and "sink" tissues may be coordinated.

REGULATION OF SUCROSE SYNTHESIS

<u>Metabolic Fine Control</u>. Sucrose synthesis in mesophyll cells requires the coordination of CO_2 fixation in the chloroplast and sucrose formation in the cytoplasm. This coordination insures that metabolite depletion within the plastid does not occur when conditions for CO_2 fixation are suboptimal. Regulation of the sucrose formation pathway is required for this coordination, and also appears to play an important role in regulating assimilate export from source leaves.

The regulation of sucrose formation is complex and involves several mechanisms. Cytoplasmic fructose-1,6-bisphosphate-1,phosphatase (FbPase) and sucrose phosphate synthase (SPSase) are thought to be important control points as the activity of both enzymes appears to be regulated metabolically in vivo (5, 7, 8). Metabolites that may regulate cytoplasmic FbPase activity in vivo include fructose-1,6-bisphosphate (Fru-1,6-P_2), AMP, dihydroxy acetone phosphate (DHAP), and the regulatory metabolite fructose 2,6-bisphosphate (Fru-2,6-P_2) (4, 7, 8, 21). The activity of SPSase is regulated by glucose 6-phosphate (Glc-6-P) and Pi (5). In addition, triose-phosphate (triose-P) availability may affect a coordinate metabolic regulation of cytoplasmic FbPase and SPSase. Stitt et al. (22, 23), have shown that a reduction in photosynthesis is associated with decreased cytoplasmic concentrations of DHAP and 3-phosphoglyeric acid (PGA) and increased Fru-2,6-P_2 concentrations. The rise in Fru-2,6-P_2 concentration was attributed to decreased cytoplasmic triose-P concentration because DHAP and PGA inhibit fructose 6-P,2-kinase, the enzyme involved in Fru-2,6-P_2 synthesis (2). Cytoplasmic FbPase activity would be reduced in vivo as a result of increased concentrations of Fru-2,6-P_2, which is a potent inhibitor of FbPase (4, 8, 24). Inhibition of FbPase will lead to decreased production of hexose phosphates such as Glc-6-P. Because Glc-6-P is an activator of SPSase (5), a reduced hexose-P pool would result in lower SPSase activity in vivo. Thus, triose-P availability may influence sucrose formation by affecting a coordinate regulation of cytoplasmic FbPase and SPSase.

Although the rate of sucrose formation decreases as photosynthetic rate is restricted, it is not necessarily true that the rate of sucrose formation will increase when net carbon exchange (NCE) rate is increased above that observed under normal conditions (high light, ambient CO_2). For example, CO_2 enrichment experiments with soybean (16) and tomato (9) suggest that sucrose formation and export do not increase in parallel with NCE rate when plants are initially exposed to atmospheres enriched with CO_2. Results such as these suggest that biochemical factors within the leaf cell, in addition to NCE rate, may influence the capacity for sucrose synthesis.

Coarse Control of Sucrose Formation. Although the sucrose formation pathway is subject to metabolic fine control, the pathway also appears to be regulated by coarse control of SPSase activity. SPSase activity measured in vitro is often just sufficient to account for sucrose synthesis in vivo, and may reflect the capacity of the pathway under optimal conditions for photosynthesis (6). The activity of SPSase in leaves has been reported to vary among species (11), and SPSase activity was negatively correlated with partitioning of carbon into starch. Because the partitioning of fixed carbon into starch and sucrose are often reciprocally related (10, 13, 15), it was not unexpected that SPSase activity and assimilate export rate would be positively correlated. Figure 1 compares export rate and SPSase activity in leaves of several species. Over all the data, a clear positive relationship is evident. In addition to genotypic variation, growth conditions also affected both parameters. The most dramatic difference was noted between tobacco plants grown in the greenhouse and phytotron. Plants in the phytotron had a higher export rate and SPSase activity than those in the greenhouse, while NCE rate was not affected (Fig. 1, inset). Although there was substantial variation in NCE rate, the differences in assimilation rate could not entirely account for the range in export rates observed. The results suggest that SPSase activity, in addition to NCE rate, may influence sucrose formation and export. Other studies support the postulate that SPSase activity often reflects the "capacity" for sucrose synthesis. In general, higher SPSase activity has been associated with increased partitioning of carbon into sucrose, increased translocation of assimilates from leaves, and decreased formation of starch (10, 15, 17). Collectively, the results suggest that metabolic fine control of cytoplasmic FbPase and SPSase along with coarse control of the activity of SPSase may serve to regulate sucrose synthesis in leaves.

<u>Diurnal Rhythm in SPSase Activity</u>. In addition to being controlled by genetic and environmental factors, we have observed that SPSase activity in leaves of 'Ransom' soybean plants fluctuates diurnally (18, 20), while the activities of other enzymes involved in sucrose metabolism do not (20). The rhythm in SPSase activity persisted under continuous conditions and appears to be controlled by an endogenous clock mechanism (18). The diurnal pattern of SPSase activity in vegetative soybean plants often exhibits a bimodal pattern--one peak occurs in the first part of the light period followed by a second peak about 12 h later in the dark. The timing and amplitude of each peak varies among experiments, as is often observed in endogenous rhythms. To assess the physiological significance of

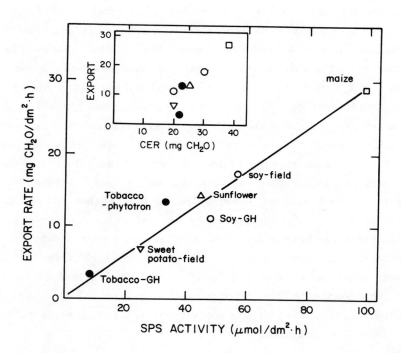

Fig. 1. Relationship between export rate and SPSase activity in fully expanded leaves of several species. The inset compares export rate with NCE rate. Measurements of photosynthesis and mass carbon export (24) were made during a 4-h interval about midday. At the end of the interval, leaves were harvested for enzyme assays (16). Soybeans (soy) were grown in the field or greenhouse (GH) as indicated. Tobacco plants were grown in pots in the phytotron or greenhouse (GH). Values are means of at least four measurements. SPSase activity noted as SPS activity, CER is NCE rate.

diurnal fluctuations in SPSase activity to photosynthetic carbon metabolism, we have measured NCE rate, assimilate export rate, and SPSase activity in field-grown soybean plants at two different stages of development. Pronounced diurnal fluctuations in SPSase activity have been observed in soybean 'Davis' plants in late vegetative development (Fig. 2A) and at the stage of midpodfill (Fig. 2C). The pattern observed during vegetative growth (Fig. 2A) is generally similar to the "typical" soybean rhythm: a peak in the first hours of the photoperiod followed by a second peak into the beginning of the dark period. Figure 2B shows changes in certain photosynthetic parameters in mature leaves of vegetative plants over the 24-h period. NCE rate was highest in the morning, and then progressively declined. Assimilate export rate declined with NCE rate throughout most of the light period, but then increased just prior to sunset. Over the course of the photoperiod, lowest export rates occurred between 1400 and 1700 h, which corresponded with the lowest SPSase activities.

Different patterns were observed when plants were in reproductive development (Fig. 2C). A single peak of SPSase activity occurred during the second half of the light period (Fig. 2C). NCE rate and export increased until midday and then decreased during the remainder of the light period (Fig. 2D). The basis for the different patterns observed in vegetative versus reproductive plants is not clear at present. Conceivably, stage of development or changes in environmental conditions that occur during the season may be involved.

As noted above, changes in the SPSase activity and export are often similar during the day. In the dark, export rate is not related to SPSase activity (19, 20) and appears to be controlled by other factors such as the rate of starch breakdown. Figure 3 compares the in vivo export rate with the in vitro SPSase activity from five separate experiments conducted in the greenhouse with vegetative soybean plants. The theoretical line (dashed line in Fig. 3) gives the maximum rate of export (assuming all exported assimilate is sucrose) that could be supported by a given activity of SPSase. All the experimental lines fell to the right of the theoretical line, which may reflect that conditions (e.g. substrate concentrations) for SPSase activity were more nearly optimal in vitro than in vivo. The results indicate that export rate and SPSase activity vary in parallel,

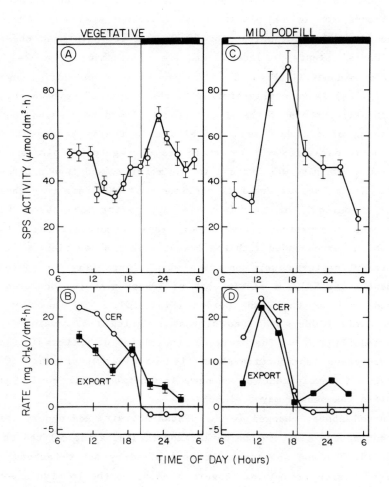

FIG. 2. Diurnal changes in SPSase activity (A, C) and photosynthetic parameters (B, D) of soybean leaves measured during vegetative (left panel) or reproductive (right panel) growth. Plants were grown in an irrigated field to eliminate growth constraints (pot effects). Values are means of six or 12 replicate determinations in A and C, and B and D, respectively. Photoperiod is indicated by the bar at the top of the figure. SPSase activity noted as SPS Activity, CER is NCE rate.

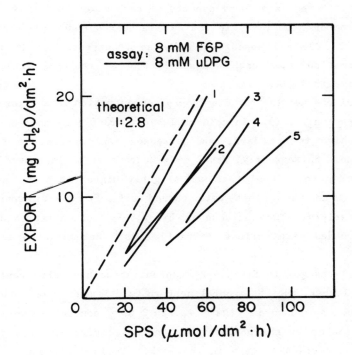

FIG. 3. Relationship between diurnal changes in export rate and SPSase activity in soybean leaves from five separate greenhouse experiments. Each line was obtained by linear regression (significant at the 0.05 level) of values obtained at different times of the day. The dashed line is the theoretical maximum export rate that could be supported for a given SPSase activity. SPSase activity noted as SPS.

and although a cause-and-effect relationship is not proven, the consistent association suggests that the diurnal rhythm of SPSase activity may contribute to changes in export rate during the day. It is important to note that the regression lines obtained in different experiments were not identical. This suggests that other factors, such as metabolic regulation of SPSase and FbPase, may be involved in regulation of sucrose synthesis and export during the diurnal cycle.

<u>Fru-2,6-P$_2$ as a Regulatory Metabolite — Diurnal Changes.</u> Work in several laboratories has established that Fru-2,6-P$_2$ is an important metabolite that regulates many aspects of carbon metabolism (4, 24). The concentration of Fru-2,6-P$_2$ in leaves fluctuates under different conditions (3, 22, 23) and appears to play an important role in regulating sucrose formation when carbon assimilation is limiting (22), or when export is impaired (14, 23).

Diurnal changes in leaf Fru-2,6-P$_2$ concentration were first documented by Stitt <u>et al</u>. (21). In spinach leaves, Fru-2,6-P$_2$ concentration was relatively high in the dark, and decreased rapidly after the onset of illumination. Subsequently, Fru-2,6-P$_2$ concentration increased gradually during the day to a level that was slightly higher than that present in the dark. When the lights were turned off, Fru-2,6-P$_2$ concentration decreased rapidly. The increase in Fru-2,6-P$_2$ concentration during the day corresponded with periods of rapid starch accumulation in the leaf (21).

Diurnal changes in Fru-2,6-P$_2$ concentration have also been observed in soybean leaves, but the general pattern is different than that reported for spinach. As shown in Figure 4, Fru-2,6-P$_2$ concentration was lowest at the beginning of the photoperiod, increased progressively during the day and then decreased again in the dark. It is important to note that the experiments with soybean (Fig. 4) were conducted in the greenhouse, whereas the studies with spinach (21) were in growth chambers. The different diurnal patterns may reflect the different environmental and experimental conditions employed; however, important species differences may also exist.

Within a given experiment, the diurnal changes in Fru-2,6-P$_2$ concentration and SPSase activity are generally reciprocal (Fig. 4). This

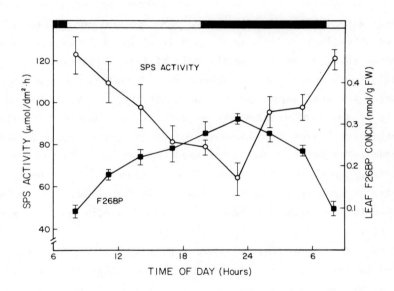

FIG. 4. Diurnal changes in SPSase activity and Fru-2,6-P_2 concentration in soybean leaves. Leaf samples were extracted and assayed for SPSase activity as described previously (16); Fru-2,6-P_2 concentration was measured using a modification of the procedure in Ref. 15. Recovery of Fru-2,6-P_2 remained constant at about 50%; values are not corrected for recovery. SPSase activity noted as SPS activity, F26BP is Fru-2,6-P_2.

inverse relationship suggests that Fru-2,6-P_2 concentration and SPSase activity are coordinated factors that may act to regulate sucrose formation. It is possible that the reduction in SPSase activity may lead to an accumulation of phosphorylated intermediates (especially F6P) which may then affect Fru-2,6-P_2 synthesis/degradation (23).

Diurnal Fluctuations in Growth of Plant Parts. It is generally recognized that growth of plant parts is ultimately dependent upon assimilates imported from source leaves. Since the rate of assimilate export from leaves changes diurnally, we are interested in whether rhythm in growth may also occur, and if so, their relation to assimilate availability.

Numerous studies have found that turgor pressure is correlated positively with leaf elongation rate. However, Takami et al. (25), have suggested other factors such as the supply of assimilates may also play a significant role in controlling the rate of leaf expansion.

Leaf growth can occur throughout the diurnal cycle. In the light, assimilates for leaf expansion can be derived from the concurrent photosynthesis of the immature leaves, as well as from sucrose imported from adjacent source leaves. In the dark, continued growth would be dependent upon accumulated carbon reserves or imported sucrose. In vegetative soybean plants, synthesis of sucrose from starch mobilization in source leaves is the primary source of carbon in the dark, and stems and expanding leaves in the shoot are the major sites of growth and utilization of carbon (12).

It has been recognized for some time that the diurnal rate of leaf expansion is not constant. In the 1950's, Bünning (1) reported oscillations in soybean leaf expansion rate during the light/dark cycle, and found that leaf growth continued for 3 to 5 d under constant environmental conditions. In similar experiments with soybeans, we observed that leaf growth essentially stops after the end of the "normal" dark period, and was coincident with depletion of carbohydrate reserves in the subtending source leaves (12). In addition, leaf growth at different nodes was synchronized, and rhythmic growth persisted in continuous light (manuscript in preparation). We have recently observed that elongation of other plant parts (stem and petiole) is also rhythmic, and the elongation of parts at different positions is synchronized. Figure 5 compares diurnal changes in growth of leaves, petioles, and stems in soybean plants. For each part, growth at two positions was summed and plotted. At the beginning of the light and dark periods, all parts were rapidly growing and thus competing for available assimilates. However, at other times, rhythms of growth were out of phase. At the end of the day, leaf growth predominated, whereas at the end of the night, petiole and stem growth was favored. Estimated "total" (leaves, stems, and petioles) growth tended to decline both during the day and at night in a pattern that was qualitatively similar to that of assimilate export from the subtending source leaves. We tentatively postulate that dry matter allocation among plant parts may reflect the interaction between endogenous rhythms in assimilate export from "source" leaves and growth of individual plant parts.

The biochemical basis for the apparent endogenous rhythm in leaf expansion is poorly understood. Carbon reserves in expanding sink leaves are relatively low (12) and growth may be dependent upon concurrent import and metabolism of assimilates. It is interesting to note that diurnal fluctuations in leaf growth are closely paralleled by changes in sink leaf Fru-2,6-P_2 concentration (Fig. 6). The changes in Fru-2,6-P_2 concentration appeared to precede slightly the changes in leaf expansion, and thus, expansion rate was correlated positively with Fru-2,6-P_2 concentration at the beginning of the respective time interval during which growth was measured (Fig. 6, inset). Diurnal changes in sink leaf Fru-2,6-P_2 concentration conceivably would affect the glycolytic utilization of imported assimilates (4) which, in turn, may play a role in regulating leaf growth. Many questions remain to be answered; however, the results suggest that Fru-2,6-P_2 plays an important role in plant metabolism in both source and sink leaves.

CONCLUSION

The rate of assimilate export from a recently fully expanded soybean leaf fluctuates during the day, as does the activity of SPSase--a key enzyme of the sucrose formation pathway. In many separate experiments, the two parameters change in parallel and thus coarse control of SPSase

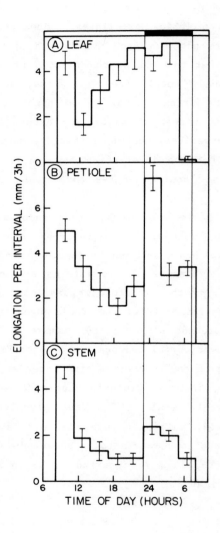

FIG. 5. Diurnal changes in the rate of (A) leaf, (B) petiole, and (C) stem elongation in vegetative soybean plants. Values are means ± SE (N = 6) from nondestructive measurements made on the same plants.

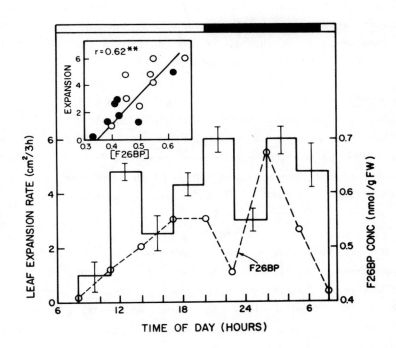

FIG. 6. Diurnal changes in the rate of expansion and Fru-2,6-P_2 concentration in expanding soybean leaves. The inset compares expansion rate (during a 3-h interval) with the concentration of Fru-2,6-P_2 (shown as F26BP) present at the beginning of each interval. Each value is the mean of four determinations. The open and closed symbols in the inset are from two separate experiments.

activity, in addition to metabolic control, may regulate sucrose formation and assimilate export. Leaf Fru-2,6-P_2 concentration fluctuates diurnally and the changes are reciprocal to those in SPSase activity. We postulate that diurnal changes in SPSase activity and Fru-2,6-P_2 concentration coordinately control the rate of carbon flux into sucrose. In general, the rate of assimilate export from leaves of vegetative soybean plants tends to decrease during the day but increases during the day in leaves of reproductive plants. Thus, the timing and supply of assimilates from mature leaves may be affected by developmental and/or environmental factors.

Growth of aerial plant parts also fluctuates diurnally and parts at different positions are generally synchronized, but not always in phase. The relations among these rhythms may be one of the factors that determines dry matter allocation among plant parts.

Acknowledgments--Cooperative investigations of the Agricultural Research Service, United States Department of Agriculture, and the North Carolina Agricultural Research Service, Raleigh, NC. Paper of the Journal series of the North Carolina Agricultural Research Service, Raleigh, NC.

A number of colleagues have contributed to the ideas and data presented, including M. Bickett, T. W. Rufty, and D. Doehlert.

LITERATURE CITED

1. BUNNING E 1956 Leaf growth under constant conditions and as influenced by light-dark cycles. In FL Milthorpe, ed, The Growth of Leaves, Butterworths Sci, London, pp 119-126
2. CSEKE C, BB BUCHANAN 1983 An enzyme synthesizing fructose 2,6-bisphosphate occurs in leaves and is regulated by metabolite effectors. FEBS Lett 155:139-142
3. CSEKE C, M STITT, A BALOGH, BB BUCHANAN 1983 A product-regulated fructose 2,6-bisphosphatase occurs in green leaves. FEBS Lett 162:103-106
4. CSEKE C, NF WEEDEN, BB BUCHANAN, K UYEDA 1982 A special fructose bisphosphate functions as a cytoplasmic regulatory metabolite in green leaves. Proc Natl Acad Sci 79:4322-4326

5. DOEHLERT DC, SC HUBER 1983 Regulation of spinach leaf sucrose phosphate synthase by glucose-6-phosphate, inorganic phosphate, and pH. Plant Physiol 73:989-994
6. FOYER CH, S HARBRON, D WALKER 1981 Regulation of sucrose phosphate synthase and sucrose biosynthesis in spinach leaves. G Akoyunoglou, ed, Photosynthesis IV. Regulation of Carbon Metabolism, Balaban Int Sci Service, Philadelphia, PA, pp 357-364
7. HARBRON S, C FOYER, DA WALKER 1981 The purification and properties of sucrose phosphate synthase from spinach leaves: The involvement of this enzyme and fructose bisphosphatase in the regulation of sucrose biosynthesis. Arch Biochem Biophys 212:237-246
8. HERZOG G, M STITT, HW HELDT 1984 Control of photosynthetic sucrose synthesis by fructose 2,6-bisphosphate. III. Properties of the cytosolic fructose 1,6-bisphosphatase. Plant Physiol 75:561-565
9. HO LC 1978 The regulation of carbon transport and the carbon balance of mature tomato leaves. Ann Bot 42:155-164
10. HUBER SC 1981 Inter- and intra-specific variation in photosynthetic formation of starch and sucrose. Z Pflanzenphysiol 101:49-54
11. HUBER SC 1981 Interspecific variation in activity and regulation of leaf sucrose phosphate synthase. Z Pflanzenphysiol 102:443-450
12. HUBER SC 1983 Relation between photosynthetic starch formation and dry weight partitioning between the shoot and root. Can J Bot 61: 2709-2716
13. HUBER SC 1983 Role of sucrose phosphate synthase in partitioning of carbon in leaves. Plant Physiol 71:818-821
14. HUBER SC, DM BICKETT 1984 Evidence for control of carbon partitioning by fructose-2,6-bisphosphate in spinach leaves. Plant Physiol 74: 445-447
15. HUBER SC, DW ISRAEL 1982 Biochemical basis for partitioning of photosynthetically fixed carbon between starch and sucrose in soybean [Glycine max (L.) Merr] leaves. Plant Physiol 69:691-696
16. HUBER SC, HH ROGERS, DW ISRAEL 1984 Effects of CO_2 enrichment on photosynthesis and photosynthate partitioning in soybean [Glycine max (L.) Merr] leaves. Physiol Plant 62:95-101

17. KERR PS, SC HUBER, DW ISRAEL 1984 Effect of N-source on soybean leaf sucrose phosphate synthase, starch formation, and whole plant growth. Plant Physiol 75:483-488
18. KERR PS, TW RUFTY JR, SC HUBER 1984 Endogenous rhythms in photosynthesis, sucrose phosphate synthase activity and stomatal resistance in leaves of soybean [Glycine max (L.) Merr]. Plant Physiol. In press
19. RUFTY TW JR, SC HUBER, PS KERR 1984 Effects of canopy defoliation in the dark on starch mobilization and activities of sucrose biosynthetic enzymes. Plant Sci Lett 34:247-252
20. RUFTY TW JR, PS KERR, SC HUBER 1983 Characterization of diurnal changes in activities of enzymes involved in sucrose biosynthesis. Plant Physiol 73:428-433
21. STITT M, R GERHARDT, B KURGEL, HW HELDT 1983 A role for fructose-2,6 bisphosphate in the regulation of sucrose synthesis in spinach leaves. Plant Physiol 72:1139-1141
22. STITT M, B HERZOG, HW HELDT 1984 Control of photosynthetic sucrose synthesis by fructose 2,6-bisphosphate. I. Coordination of CO_2 fixation and sucrose synthesis. Plant Physiol 75:548-553
23. STITT M, B KURZEL, HW HELDT 1984 Control of photosynthetic sucrose synthesis by fructose 2,6-bisphosphate. II. Partitioning between sucrose and starch. Plant Physiol 75:554-560
24. STITT M, G MIESKES, HD SOLING, HW HELDT 1982 On a possible role of fructose 2,6-bisphosphate in regulating photosynthetic metabolism in leaves. FEBS Lett 145:217-222
25. TAKAMI S, HM RAWSON, NC TURNER 1982 Leaf expansion of four sunflower (Helianthus annuus L.) cultivars in relation to water deficits. II. Diurnal patterns during stress and recovery. Plant Cell Environ 5:279-286

The Influence of External pH on Sucrose Uptake and Release in the Maize Scutellum

THOMAS E. HUMPREYS

Slices of the maize scutellum accumulate sucrose when incubated with fructose, glucose, or sucrose (3). The plasmalemma of the scutellum cell is freely permeable to hexoses and so hexoses are not accumulated, whereas sucrose uptake is an active process that appears to involve a sucrose-H^+ cotransport (symport) system (5, 8). Briefly, the evidence for a sucrose-H^+ cotransport system is that: (a) sucrose accumulated in scutellum slices is released when a lipid-soluble cation is added to the bathing solution (8); (b) in energy-poisoned tissue, sucrose release drives Rb^+ uptake (8); and (c) under conditions of net efflux, sucrose and H^+ are transported in a 1:1 ratio (5). The evidence in (a) and (b) indicates that sucrose transport is electrogenic and in (c) identifies the cotransported ion and gives the cotransport stoichiometry. This evidence derives from sucrose release. During sucrose uptake, the rate of acidification of the bathing solution is decreased, and immediately upon addition of sucrose to the bathing solution, there is a small and transient increase in pH (4). These results indicate that sucrose uptake involves H^+ uptake, and support the evidence derived from sucrose release. In this paper, it is assumed that sucrose uptake, like sucrose release, takes place on a cotransporter with H^+.

A cotransporter utilizes energy contained in the chemical potential gradient of H^+ ($\Delta\mu_H^+$) across a membrane to drive a net flux of sugar across the membrane. This gradient contains two components: a concentration gradient (ΔpH) and an electrical potential gradient ($\Delta\psi$). In some energy transducing systems, e.g. ATP synthesis in <u>Streptococcus lactis</u> (13), the two components are equally effective, although it is not known how the membrane machinery, e.g. the ATPase, avails itself of them. It appears that both components of $\Delta\mu_H^+$ also can be utilized by the lactose-H^+ co-transporter of Escherichia coli (9). However, research with higher plants has provided little data on the effect of $\Delta\psi$ on sugar uptake, and the data for ΔpH are conflicting (see Ref. 15 for details).

In this paper, the effects of external pH on sucrose uptake and release in maize scutellum slices are presented. The results indicate that both components of $\Delta\mu_H^+$ can drive sucrose uptake and that sucrose release is enhanced by a lowered $\Delta\psi$ and by uptake of a compensating cation. The effects of external pH on sucrose release are puzzling.

MATERIALS AND METHODS

Plant Material. Tissue slices were prepared from scutella of 3-d-old maize seedlings (Zea mays L. cv Funks 4949A) as previously decribed (6).

Incubation Conditions. Slices (0.5 g) were placed in 50-ml beakers with 10 ml of experimental solution (described below), and incubated at 30°C in a H_2O bath. The buffers (20 mM) were the Na salts of citrate (pH 5 and below), Mes (pH 5 and 6), 3-(N-Morpholino)-propane sulfonic acid (Mops) (pH 7), and N-(2-hydroxy ethyl-piperazine-N'-3-propane sulfonic acid (Epps) (pH 7.5 and 8). When required, solutions and wash H_2O were removed from the beakers with a Pasteur pipette connected to an aspirator pump.

Sucrose Uptake. Unless otherwise noted in text and figures, slices were incubated in buffer for 1 h, then washed with 10 ml H_2O, and placed in the sucrose uptake solution. The uptake solution contained buffer, 2 mM sucrose and, sometimes, 1 mM $CaCl_2$. Aliquots (0.1 ml) were removed for sucrose analysis 15 min after adding the uptake solution, and at 1 h intervals thereafter.

Sucrose Leakage. Slices were incubated in buffer for 1 h and then placed in the sucrose uptake solution. At zero time and each h after that, a group of slices was removed from the uptake solution, washed with 10 ml H_2O (2 X), and placed in buffer. Each 15 min for a 60-min period, the entire bathing solution was removed for sucrose analysis, and fresh buffer was added.

Measurement of Cytoplasmic pH and $\Delta\psi$. Cytoplasmic pH and $\Delta\psi$ were estimated from the distributions of 2,4-dinitrophenol (ΔpH) and tri-[U-^{14}C]phenyl phosphonium ions ($\Delta\psi$) between scutellum slices and bathing solution by procedures previously described (6, 7).

Sucrose Release. Groups of slices were incubated for 2 h in 100 mM fructose to increase their sucrose content, and then were placed in 50 mM mannose for 30 min. The mannose treatment drastically reduces the ATP content and inhibits the H^+ pump in the plasmalemma (1, 2). After the mannose treatment, the slices were washed with 10 ml H_2O and placed in a buffered solution containing 50 mM mannose plus or minus 0.5 mM 2,4-dinitrophenol (DNP), 20 mM triphenylmethylphosphonium ion (TPMP$^+$), or 50 mM KCl. Aliquots (0.1 ml) were removed at intervals for sucrose analysis.

RESULTS AND DISCUSSION

Sucrose Uptake and pH. Previously, it was shown that the rate of sucrose uptake in scutellum slices declined sharply as the external pH rose above 5 (4). These results were for the first 2 h of uptake. In the present experiments, however, sucrose uptake was followed over a 5-h period, and the rates of uptake at the higher pH were found to increase with time until they became about equal to the rate at pH 5 (Fig. 1). In the experiment of Figure 1, 2 mM sucrose was used, but curves of similar shape, although representing higher rates, were obtained with 5 mM sucrose. It is clear that as the pH increased from 5 to 8, the length of the initial lag in uptake increased, but the final constant rate of uptake was essentially independent of pH.

The literature contains conflicting data on effects of pH on sucrose uptake. For example, rates of sucrose uptake in aged wheat leaf fragments (11) and protoplasts isolated from developing soybean cotyledons (12) declined sharply above pH 5 or 6, whereas rates of sucrose uptake in whole immature soybean embryos (16) and detached whole cotyledons of castor bean (10) were little influenced by pH. It appears that pH sensitivity is associated with tissues that may have been injured in preparation, e.g. sliced wheat leaves, isolated protoplasts, or the thinly sliced scutella used for Figure 1.

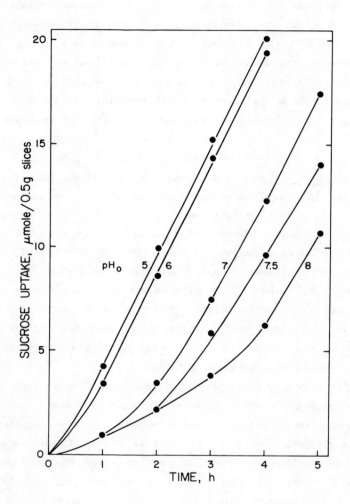

FIG. 1. Effect of external pH on sucrose uptake. Curves have been corrected for sucrose leakage, see text.

Leakage, Calcium, and the Lag Period. One possible cause of the lag period, a pH-dependent leak which stops with time, was eliminated by correcting the uptake data for leakage. After this correction is applied to the curves for Figure 1, the lag period remains.

Sucrose leakage was independent of pH in the 5 to 7 range, but increased about 20% at pH 8 (data not shown). Leakage rates at pH 7 as a function of time of incubation in the sucrose uptake solution are shown in the bottom curve of Figure 2. Rates of leakage decreased with time and reached zero during the last h of incubation. Therefore, in Figure 1, only the shape of the curves during the lag period and not the final steady rate of uptake was influenced by the applied leakage corrections. This is taken to mean that leakage represents a passive movement (through a water-filled pore in the plasmalemma, perhaps), whereas uptake represents a net influx on the sucrose-H^+ cotransporter. In making the leakage corrections, it was assumed that the measured leakage rate was the same as that occurring during sucrose uptake.

Calcium (1 mM) strongly inhibited leakage (Fig. 2, top curve). Therefore, if leakage caused the lag, Ca^{2+} should shorten the lag period. Figure 3 shows the effect of Ca^{2+} on sucrose uptake at pH 7, and the effects of leakage corrections on the shapes of the curves. Calcium increased the lag period and inhibited sucrose uptake 19%. The Ca^{2+} results support the above conclusion that the lag is not a result of a pH-dependent leak.

In some experiments, 10 mM Ca^{2+} was used, and sucrose uptake was inhibited about 60% at pH 7 (data not shown). The Ca^{2+} inhibition was pH-dependent, increasing as the pH increased (4), and this may account for some of the results in the literature showing a pH dependency for sucrose uptake. For example, with protoplasts from soybean cotyledons, sucrose uptake at pH 8 was only 25% of that at pH 5, and the uptake solution contained 10 mM Ca^{2+} (12).

Preincubation and the Lag Period. If the lag is caused by injury, then the results of Figure 1 indicate either that recovery from injury is pH dependent, being very rapid at pH 5 and slow at pH 8, or that injury is pH dependent, being caused by a pH above 5. To check these ideas, slices were preincubated in buffer or H_2O before being placed in the

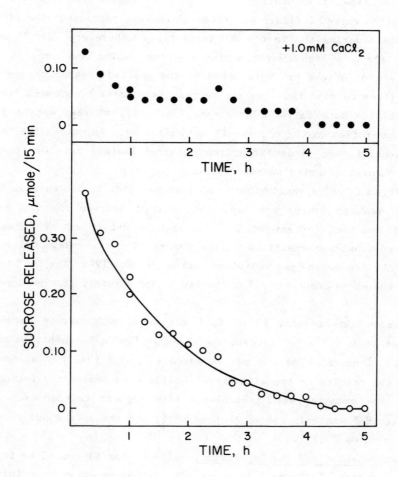

FIG. 2. Sucrose release (leakage) at pH 7 in the presence (upper curve) or absence (lower curve) of 1 mM CaCl$_2$.

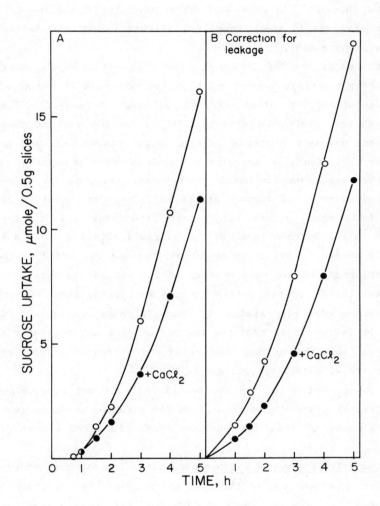

FIG. 3. Effect of $CaCl_2$ on sucrose uptake at pH 7. Sucrose leakage corrections (from Fig. 2) were applied to the data of A to give the curves in B.

uptake solution at pH 7. As the length of the preincubation period in buffer increased, the length of the lag period decreased, but the final steady rate of uptake was not affected (Fig. 4). When preincubation was in H_2O (pH varied from ca. 5.5-4.5 during preincubation), results very similar to those of Figure 4 were obtained. It is concluded that, if injury is involved in the uptake lag, neither injury nor recovery from injury is pH-dependent.

In the experiment of Figure 4, 2 mM EDTA was added to one group of slices during the first hour of a 2-h preincubation. EDTA had no effect on the subsequent sucrose uptake (Fig. 4), although it removed Mg^{2+} and Ca^{2+} from the slices. This indicates that the lag period was not caused by a pH-dependent movement of these cations, e.g. between wall and membrane.

Komor (10) found a correlation between sucrose uptake rate and internal sucrose concentration in castor bean cotyledons. He concluded that increased rates of sucrose uptake following preincubation reflect a decrease in internal sucrose, not recovery from injury. In the experiment of Figure 1, the sucrose level of the slices declined ca. 50% during the 5-h uptake period. However, when 10 mM fructose was added to the 2 mM sucrose solution, the sucrose content of the slices was maintained above the initial level for the entire 5-h uptake period, and the curves for sucrose uptake were very similar in shape to those of Figure 1 (data not shown). It is concluded that the lag period does not represent the time required to lower the sucrose content of the slices to a level that does not interfere with sucrose uptake.

The cause of the lag has not been found. It may involve $\Delta\psi$, which makes a greater contribution to $\Delta\mu_H^+$ as the external pH is increased (7). The significance of this for sucrose uptake is discussed in the next section.

<u>Protonmotive Force and Sucrose Uptake</u>. The net rate of sucrose-H^+ cotransport is a function of the chemical potential gradients of both H^+ and sucrose. In the experiments of Figure 1, the sucrose gradients probably were independent of external pH, and therefore, rates of uptake in the 5 to 8 pH range should be a function solely of $\Delta\mu_H^+$. The results of Figure 5, where rates of sucrose uptake (from Fig. 1) and Δp ($\Delta\mu_H^+/F$, where F is Faraday's constant) are plotted against external pH, are

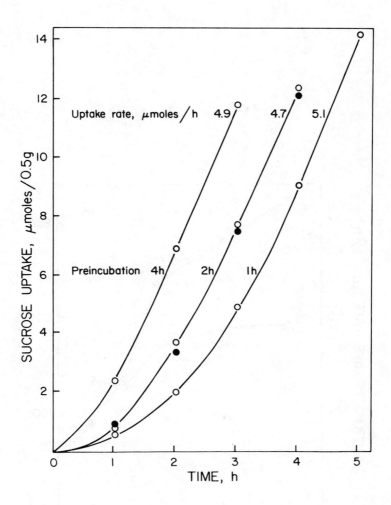

FIG. 4. Effect of length of preincubation on sucrose uptake at pH 7. Slices were preincubated at pH 7 for 1, 2 or 4 h before being placed (time zero) in the sucrose uptake solution. One group of slices was incubated in buffered 2 mM EDTA (pH 7) for 1 h, followed by buffer alone for 1 h before being placed in the uptake solution (closed circles).

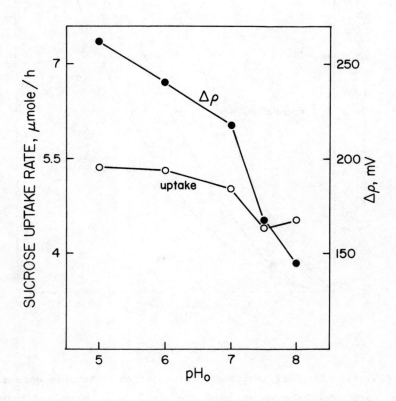

FIG. 5. Effect of external pH on the rate of sucrose uptake and on the protonmotive force (Δρ, see Text).

largely in agreement with the above statement, considering that the uptake rate will be linearly related to Δp only in near-equilibrium situations, and that the cotransporter might be unable to utilize Δp to an equal extent over the entire pH range.

At higher pH, Δp was made up largely of $\Delta \psi$, whereas at pH 5, Δ pH predominated (Fig. 6). Therefore, it appears that sucrose uptake can be driven by either component of Δp. However, this does not mean that the two driving potentials are equivalent per mvolt, and the results of Figure 6 do not contain this information. The near constant rate of sucrose uptake with wide variations in $\Delta \psi$ and Δ pH could mean that Δp was not rate limiting. Nevertheless, it is clear that $\Delta \psi$ alone can drive sucrose uptake in the maize scutellum (Fig. 6).

<u>The Proton Well</u>. To account for the equivalence of the electrical and chemical components of $\Delta \mu_{H^+}$ in his chemiosmotic hypothesis of ATP synthesis, Mitchell (14) introduced the idea of a "proton well" in the ATPase of the energy-transducing membrane. Protons, in moving down this well, move down the electric field of the membrane, and the resulting decrease in the contribution of $\Delta \psi$ to μ_{H^+} is balanced by an increase in H^+ activity. The chemical potential gradient everywhere within the well is equal to $\Delta \mu_{H^+}$ in the bulk solution outside the well, but the relative contributions of $\Delta \psi$ and Δ pH to $\Delta \mu_{H^+}$ differ with position in the well.

The idea of a proton well is useful in discussing sucrose-H^+ cotransport. In Table I, the pH at the bottom of the well has been calculated from the external pH and $\Delta \psi$ (Fig. 6), assuming that the bottom of the well penetrates to the inner surface of the membrane. The assumption is that the entire $\Delta \psi$ is available to drive transport and, therefore, that the H^+ binding site of the cotransporter is situated (for influx) just inside the inner surface of the membrane. In Table I, Δ pH differences of 2 units resulted in less than a 20% difference in uptake rate. If the pK of the active site was 6.5 or a little less, decreases in the pH at the bottom of the well below 5.5 would result in only a small increase in the amount of protonated cotransporter available for sucrose transport. On the other hand, if the active site had a pK of 6.5, probably only 10% or less of the cotransporter molecules facing the cytosol would be proto-

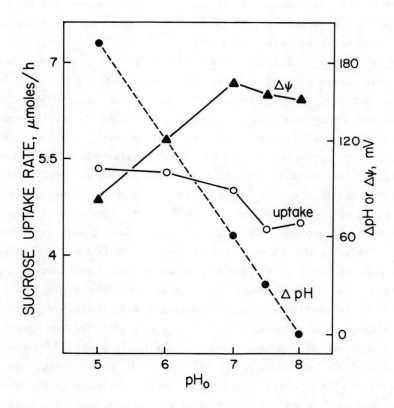

FIG. 6. Effect of external pH on the rate of sucrose uptake and on the electric potential and pH gradients across the plasmalemma.

Table I. Calculated pH at the Bottom of the Proton Well and the Resulting Δ pH.

pH_o (bulk solution)	Δψ (mv)	pH^a (proton well)	pH_i (cytosol)	ΔpH	Uptake (μmol h^{-1})
5.0	82	3.6	8.0	4.4	5.3
6.0	121	4.0	8.0	4.0	5.2
7.0	155	4.4	8.0	3.6	5.0
7.5	148	5.0	8.0	3.0	4.4
8.0	145	5.6	8.0	2.4	4.5

[a] Calculated from the data of Figure 6, using the Equation: $pH_{well} = pH_o - \Delta\psi(F/2.3\ RT)$.

nated. Therefore, in the absence of influx, sucrose efflux should be very slow (see below).

Sucrose Release. Using proteoliposomes containing lactose carrier protein from E. coli, Kaback and associates (9) showed that lactose efflux was enhanced by ionophores that collapse Δψ and by a high outside pH. The release of sucrose from scutellum cells also is enhanced under these conditions (5), and the model proposed for sucrose-H$^+$ cotransport in the scutellum (4) follows that proposed for lactose-H$^+$ cotransport in E. coli which has a firm experimental base (9). Briefly, it is proposed that (1) the scutellum cotransporter carries a negative charge, which is neutralized upon protonation; (2) the cotransporter can move across the membrane in either direction only when it is complexed with both sucrose and H$^+$, or when it is uncomplexed; and (3) during efflux, sucrose is released at the outer face of the membrane before H$^+$. The explanation for the enhancement of efflux by ionophores and high outside pH is that both increase the number of unpronated cotransporter molecules at the outside face of the membrane, and thereby, increase the rate of return of cotransporter to

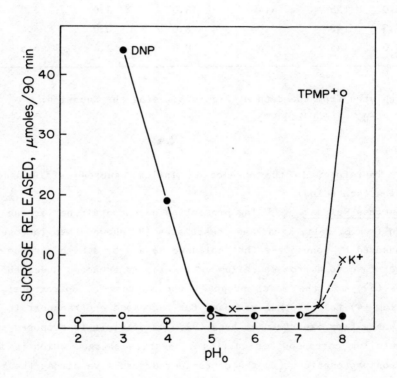

FIG. 7. Effects of a protonophore (DNP) and of cations (K^+ and $TPMP^+$) on sucrose release from scutellum slices as a function of external pH. The small amounts of sucrose released into buffer alone have been subtracted from each data point.

the inside face. Additionally, since the cotransporter has a negative charge, an ionophore may enhance efflux by collapsing $\Delta\psi$, against which the cotransporter must move when returning to the inside face.

The effect of outside pH on the release of sucrose from scutellum slices in the presence of K^+, $TPMP^+$, or DNP is shown in Figure 7. $TPMP^+$ (a rapidly penetrating, lipid-soluble cation) and DNP (a protonophore) should collapse (2, 5), and K^+ (7) should lower $\Delta\psi$. With $TPMP^+$ and K^+, there was a relatively great increase in sucrose release above pH 7, indicating perhaps that at lower pH, most of the outside-facing cotransporter molecules were protonated and could not recycle to the inner surface (cf. 9). With DNP, however, little sucrose was released until the pH fell below 5, where DNP is a good protonophore. How does the cotransporter recycle when the pH is low? The rate of sucrose release in DNP at pH 4 was doubled upon addition of $TPMP^+$ (data not shown), which indicates that sucrose release was electrogenic, i.e., that there was a net release of H^+. Unfortunately, these results cannot be explained with the present data.

LITERATURE CITED

1. GARRARD L, T HUMPHREYS 1969 The effect of mannose on sucrose storage in the corn scutellem: evidence for two sucrose transport mechanisms. Phytochemistry 8:1065-1077
2. HUMPHREYS T 1975 Dinitrophenol-induced hydrogen-ion influx into the maize scutellum. Planta 127:1-10
3. HUMPHREYS T 1977 Dinitrophenol-induced efflux of sucrose from maize scutellum cells. Phytochemistry 16:1359-1364
4. HUMPHREYS T 1978 A model for sucrose transport in the maize scutellum. Phytochemistry 17:679-684
5. HUMPHREYS T 1981 Sucrose-proton efflux from maize scutellum cells. Phytochemistry 20:2319-2323
6. HUMPHREYS T 1982 Cytoplasmic pH of maize scutellum cells. Phytochemistry 21:2165-2171
7. HUMPHREYS T 1983 Proton electrochemical gradients and sucrose accumulation in the maize scutellum. Phytochemistry 22:2669-2674

8. HUMPHREYS T, R SMITH 1980 Sucrose efflux from the maize scutellum. Ber Dtsch Ges 93:229-241
9. KABACK H 1983 The lac carrier protein in Escherichia coli. J Membrane Biol 76:95-112
10. KOMOR E 1977 Sucrose uptake by cotyledons of Ricinus communis L.: characteristics, mechanism, and regulation. Planta 137:119-131
11. LEGER A, S DELROT, J-L BONNEMAIN 1982 Properties of sugar uptake by wheat leaf fragments: effects of ageing and pH dependence. Physiol Veg 20:651-659
12. LIN W, M SCHMITT, W HITZ, R GIAQUINTA 1984 Sugar transport into protoplasts isolated from developing soybean cotyledons. I. Protoplast isolation and general characteristics of sugar transport. Plant Physiol 75:936-940
13. MALONEY P 1982 Energy coupling to ATP synthesis by the proton-translocating ATPase. J Membrane Biol 67:1-12
14. MITCHELL P 1968 Chemiosmotic Coupling and Energy Transduction. Glynn Research, Bodmin, England
15. REINHOLD L, A KAPLAN 1984 Membrane transport of sugars and amino acids. Annu Rev Plant Physiol 35:45-83
16. THORNE J 1982 Characterization of the active sucrose transport system of immature soybean embryos. Plant Physiol 70:953-958

Sucrose Transport: Regulation and Mechanism at the Tonoplast

ROGER WYSE, DONALD BRISKIN, and BENY ALONI

The yield of agronomic plants depends on maintaining high rates of photosynthesis for extended periods over a wide range of environmental conditions and the efficient allocation of those assimilates to the economically important component. During domestication, little change has occurred in photosynthetic rates. In fact, the light-saturated photosynthetic rate of modern cultivars is lower than their primitive ancestors (11). Thus, variability in assimilate partitioning has been an important factor in the conversion of primitive species to their high-yielding, domesticated equivalents. Therefore, an appropriate goal of plant research is to develop a basic understanding of the biochemical and molecular mechanisms controlling carbon allocation to the yield component of crop plants. The fundamental knowledge generated will form a basis for future research designed to alter allocation by genetic or chemical means.

In most plants, sucrose plays a role similar to that of glucose in animals. Sucrose is the first free sugar formed as a result of photosynthetic carbon fixation and is the major form in which carbon is transported via the phloem throughout the plant. Therefore, sucrose is the source of glucose and fructose used for energy and the carbon source for the biosynthesis of all biomass. Sucrose also acts as both a transient and long-term storage carbohydrate. Thus, understanding sucrose transport and metabolism is central to understanding the regulation of assimilate partitioning.

Role of Sinks in Allocation. Partitioning of assimilates involves the movement of fixed carbon from sources (photosynthetically active leaves or storage organs where carbohydrates are remobilized) through the phloem to sinks where the assimilates are utilized in growth or storage. Research has shown that the source determines the timing and supply of assimilates, but that competition between sinks determines allocation patterns (3, 7). Therefore, the goal of our research is to understand the biochemical mechanisms that determine the competitive ability of sinks.

Understanding the role of the sink involves studies of the entire source-transport-sink pathway. In source leaves, fixed carbon is allocated to starch, transiently stored as sucrose, or released from the mesophyll cells for export. Sink demand has been shown to effect allocation within the source leaf between export and starch accumulation. Active loading of sucrose into the phloem in source leaves increases the hydrostatic pressure in the phloem sieve element and drives mass flow through the phloem. However, the aspect of sink metabolism that regulates allocation patterns is less well understood, but must involve aspects of phloem unloading, retrieval, and storage by sink cells. It is now clear that in most sinks, sucrose unloads into the apoplastic space at the site of phloem unloading (20, 30, 32). Since flux rates through the phloem are related to the sucrose concentration gradient between source and sink, the concentration of sucrose at the site of phloem unloading will be an important determinant of the gradient. Our research approach is based on the hypothesis that the ability of a sink to compete with other sinks for limited available assimilates is determined by the ability of the sink to maintain low concentrations of sucrose at the site of phloem unloading.

Evidence that the phloem unloads into the apoplastic space is clear in reproductive sinks where no plasmodesmatal connections exist between the maternal and embryonic tissue (6, 29). Therefore, as the phloem unloads, sucrose must move through the apoplastic space prior to retrieval by the developing embryo. The pathway is less clear in vegetative and meristematic sinks. We have extensively studied sucrose movement in the storage taproot of sugar beet--a complex vegetative sink. Physiological data suggested that the phloem unloaded in the apoplast. However, the results of such experiments are always equivocal. Recently, we completed an anatomical study of the phloem unloading site and found no plasmodesmatal connections between the sieve element companion cell complex of the phloem and that of surrounding storage parenchyma cells. Thus, the anatomical data confirms that the phloem must unload sucrose into the apoplastic space in this vegetative sink. Recently, Giaquinta et al. (10) reported that in corn root tips, sucrose moved through the symplast. Therefore, meristematic sinks may be the exception to the rule of apoplastic unloading. Retrieval and storage of translocated sucrose by storage parenchyma cells may be an important determinant of sink-mobilizing ability.

Cellular Conditions Related to Transport. Attempts have been made to characterize the membrane potential and proton gradients that exist at the plasmalemma and tonoplast membranes in higher plant cells. This is a difficult process, but best estimates of conditions existing in most plant cells are depicted in Figure 1. At the plasma membrane, the potential is interior negative, and the cytoplasm has a higher pH than the external free space. Therefore, the proton motive force (PMF) is directed from outside to inside. The presence of an H^+-ATPase at the plasma membrane has been substantiated (26). Thus, gradient conditions and pump activity at the plasma membrane are consistent with the proton/sucrose co-transport model (9, 13, 16, 23).

At the tonoplast, the conditions are quite different. Here, estimates of membrane potential suggest a +10 and +20 mV positive potential in the vacuole, with respect to the cytoplasm (26). The vacuole is more acidic ($\Delta pH = \sim 2$) than the cytoplasm and, therefore, the PMF is directed from the vacuole to the cytoplasm.

Thom and Komor (28) reported Mg · ATP to be the preferred substrate for an H^+-ATPase on the outer surface of isolated vacuoles. This ATPase activity catalyzed the formation of an interior-positive membrane potential. Spanswick and his colleagues (1, 2) reported the presence of a proton-pumping ATPase on the vacuole membrane that is coupled to the movement of a compensating chloride ion. This would explain the large ΔpH, but small potential. It is clear that the conditions existing at the tonoplast are not consistent with a proton/sucrose co-transport

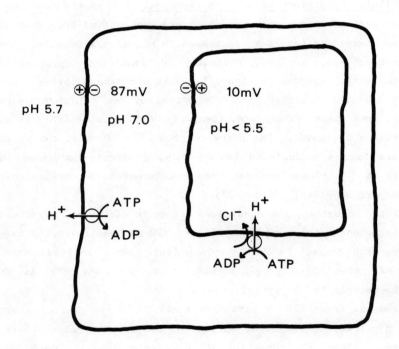

FIG. 1. Typical ΔpH and ΔΨ conditions thought to exist in parenchyma cells of higher plants. ATP-dependent pumps which can maintain ΔΨ and ΔpH are shown as circles on the membrane (heavy lines). See Text for details.

model. However, the concentration of sucrose in the vacuole of many plants may exceed 800 mM and, therefore, this transport system has important physiological implications.

Tonoplast Transport. The study of transport at the tonoplast is complicated by the inaccessibility of the membrane in tissues and cells. Only recently has the technique for isolating intact vacuoles been sufficiently developed to make it useful in transport studies (15, 27). However, the large, isolated vacuole is extremely fragile, making transport studies essentially impossible. Recently, the methodology has been developed for the isolation of sealed membrane vesicles from plant tissues (25). While sealed vesicle preparations can contain membranes derived from both the plasma membrane and tonoplast (25), these can be further resolved using density gradient centrifugation (5). Using appropriate enzyme markers, it is possible to verify the separation of plasma and tonoplast membranes (21). Transport properties in these membrane vesicles can be monitored using techniques very similar to those developed by Kaback (12) for vesicles of Escherichia coli.

In our laboratory, we have developed the technique for isolating tonoplast-enriched vesicle preparations from sugar beet taproot and for monitoring transport using fluorescent probes for $\Delta\Psi$ (membrane potential gradient) and ΔpH (4). The population of vesicles isolated would be expected to be ca. 50% inside out and 50% right side out. Those with the ATPase active site exposed on the outer surface generate $\Delta\Psi$ and ΔpH upon the addition of Mg · ATP. The components of the PMF can be determined using fluorescent dyes, such as quinacrine for ΔpH (14) and Oxonol V for $\Delta\Psi$ (24). The principle of the assay is that as these dyes are concentrated inside the vesicle (with membrane-binding changes in the case of Oxonol V), their fluorescence is quenched. This quenching may be caused by energy transfer between the probe molecules, which results in a decrease in quantum yield (14, 24).

By altering the assay condition, it is possible to alter ΔpH and $\Delta\Psi$ independently or clamp the potential at zero. In the latter case, potassium iodide and valinomycin are added. The iodide anion is extremely permeant and valinomycin is a potassium-specific ionophore. With both a permeant anion and cation present, any disequilibrium in the $\Delta\Psi$ will be compensated by the ion movement. Thus, the membrane potential is effectively clamped at zero, allowing for the regeneration of a maximum ΔpH (31).

Now, the critical question is: Will these tonoplast vesicles transport sucrose? If so, what are the characteristics of the transport system? When ΔpH is monitored with quinacrine, quenching occurs upon addition of Mg · ATP, indicating the generation of a ΔpH (high $^+$H concentration inside) (Fig. 2). When the maximum ΔpH has been generated and sucrose is added, leakage of protons out of the vesicles occurs (fluorescence increases). This leakage is sucrose specific and does not occur when mannitol or glucose are added. This result is consistent with a carrier-specific sucrose proton antiport system.

When vesicles are incubated in 10 mM [^{14}C]sucrose for periods of up to 20 min and collected on Nucleopore filters, sucrose uptake occurs in energized, but not in de-energized (plus mCl-carbonylcyanide phenylhydrazone, CCCP), vesicles (Fig. 3). The same results were observed when Mg · ATP was excluded. Therefore, a proton gradient is required for sucrose uptake.

ATP-dependent sucrose uptake followed simple saturation kinetics with an estimated K_m of 7.8 mM (Fig. 4). This simple kinetic profile is in contrast to the biphasic pattern frequently observed in tissue (19, 22; Fig. 6) and recently in protoplasts isolated from soybean cotyledons (18). These results suggest that the biphasic kinetics are not explained by transport at the tonoplast, but are either characteristic of plasma membrane transport or a complex of a two-compartment system.

When the ΔpH was dissipated with the protonophore CCCP, sucrose leaked rapidly out of the vesicles (Fig. 5). When sucrose is added to de-energized vesicles, an inside negative Ψ is generated, suggesting that when sucrose enters, a positive charge moves out (4).

These data are consistent with sucrose loading into the vacuole via a proton-sucrose antiport system; a system consistent with the known ionic gradients existing across the tonoplast. To our knowledge, this is the first direct evidence of a neutral species being accumulated by coupling to a proton antiport system.

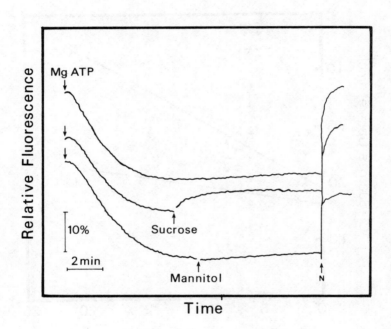

FIG. 2. Effect of sucrose and mannitol on the steady-state pH gradient in sugar beet membrane vesicles. Sugar beet tonoplast vesicles were energized by the addition of 5 mM ATP to a reaction solution containing 5 mM MgSO$_4$, 250 mM sorbitol, 50 mM KI, 10 μM valinomycin, and 100 μg of membrane protein. The H$^+$ gradient was monitored by the quenching of 5 μM quinacrine fluorescence.

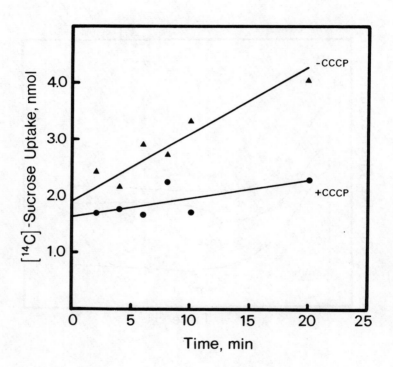

FIG. 3. Effect of CCCP on [^{14}C]sucrose uptake by energized vesicles. Vesicles were energized with Mg · ATP and then sucrose uptake monitored in the presence or absence of 5 μM CCCP. At times indicated, 50 μl aliquots of the vesicle suspension were collected by filtration and counted for radioactivity.

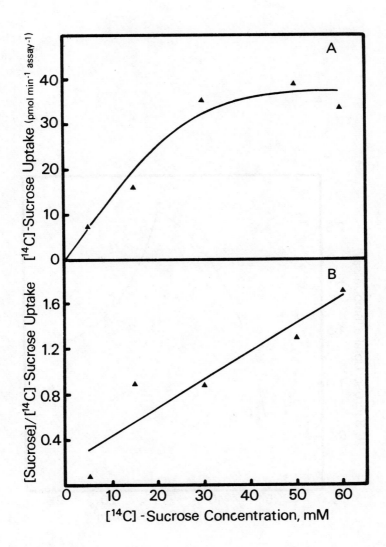

FIG. 4. Kinetics of ATP-dependent sucrose uptake by sugar beet membrane vesicles. A, Kinetic uptake profile. B, Replot using linear transformation of the Michaelis-Menton relationship.

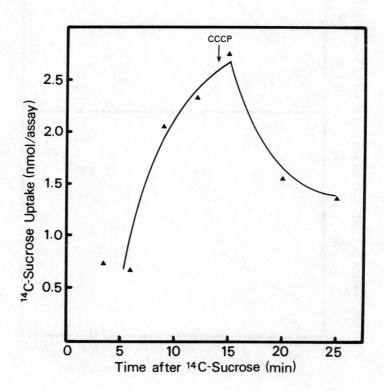

FIG. 5. Effect of CCCP on the accumulated sucrose in sugar beet tonoplast vesicles. At the point indicated, CCCP (5 μM) was added to a vesicle suspension, treated as described in Figure 4.

Sucrose Transport in Sink Tissue. Sucrose uptake by sink tissue has an interesting biphasic kinetic pattern (Fig. 6). At low external concentration, a typical Michaelis-Menton pattern is followed. But at concentrations above ca. 20 mM, the uptake increases linearly with concentration. This pattern is found not only in our taproot system (22), but also during phloem loading (19) and sucrose transport into soybean cotyledons (17). Recently, Lin et al. (18) have shown the same pattern for sucrose uptake in protoplasts isolated from developing soybean cotyledons. Therefore, this biphasic pattern is a characteristic of the transport system and not an artifact of diffusion in tissue pieces. The saturating component has been proposed to be a proton co-transport system (19), but the linear component still defies explanation. Recently, we have spent considerable time studying the regulation of sucrose transport in tissue with particular interest in the regulation of the two phases.

Turgor Regulation. Tissues which store osmotically-active compounds, such as sucrose, must be able to regulate turgor in order to prevent excessive cell turgor. The sugar beet taproot is a good, if extreme, example of such a sink. During the growing season, the osmotic concentration of the cell sap commonly increases to the equivalent of 20 bars solely as a result of sucrose accumulation. Therefore, the tissue must turgor regulate, or the turgor pressure within a cell would exceed 20 bars at maturity. This does not occur. Thus, it is logical to hypothesize that turgor may have an effect on sucrose transport systems in this sink.

We have determined the kinetics of sucrose uptake in taproot tissue discs equilibrated in various external osmotica ranging from 0 to 600 milliosmolar. The tissue utilized in these studies had a cell sap concentration in excess of 700 milliosmolar; therefore, positive turgor was maintained in the tissue under all treatment conditions.

When the tissue was equilibrated in buffer, the uptake of sucrose increased in a linear fashion with increasing concentration (Fig. 7). As the external osmotic concentration was increased (and the turgor of the cells decreased), uptake not only increased, but the saturating component of the kinetic profile increased substantially. These results suggested

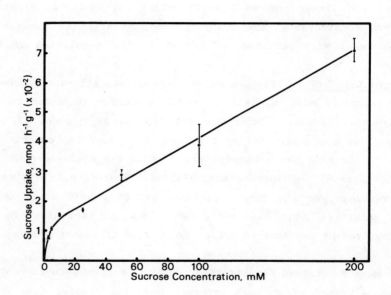

FIG. 6. Kinetics of sucrose uptake by sugar beet taproot tissue discs. Tissue discs were equilibrated in 250 mM mannitol for 90 min prior to incubation in [^{14}C]sucrose for 1 h. Mannitol and sucrose concentrations were balanced to maintain a constant 250 milliosmolar concentration. At the end of the uptake period, the tissue was rinsed in 250 mM mannitol (3 x 3 min) before the tissue was killed in hot 80% ethanol and ^{14}C determined by liquid scintillation counting.

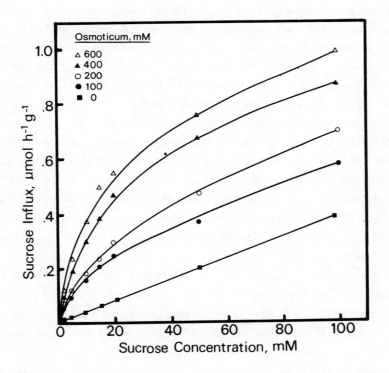

FIG. 7. Kinetic profile of sucrose uptake at various external mannitol concentrations (osmoticum = mannitol + sucrose, see Text).

that the saturating component was very sensitive to cell turgor. When the data are plotted using an Eadie-Hofstee transformation, the two components of uptake become more obvious (Fig. 8). The effect of increasing turgor was to reduce V_{max} from 590 to 290 nmol g^{-1} fresh weight h^{-1} with little change in the K_m (12.5 versus 9.0 mM).

The effect of turgor appeared to be specific for sucrose since glucose uptake exhibited saturation-type kinetics at all external osmotic concentrations (Fig. 9), although the V_{max} increased 50% when cell turgor was reduced. This must be contrasted with a total elimination of the saturating component for sucrose uptake at high turgor.

The water potential of plant cells consists of two components—an osmotic and a turgor component. To differentiate between the roles of the two components on transport characteristics, we compared a nonpermeant osmoticum (mannitol) to a permeant osmoticum (ethylene glycol). The results are given in Figure 10. The results clearly show that mannitol greatly enhances, but ethylene glycol has no effect on, the saturating component. Thus, the effect of mannitol in the external solution is an effect of turgor on the sucrose transport system.

Para-chloro-mercuribenzoate sulfonate (PCMBS) is a nonpermeant sulfhydryl inhibitor which, under the proper conditions, has been shown to inhibit the sucrose carrier, but not effect cellular metabolism (8). Maynard and Lucas (19) also demonstrated that PCMBS would inhibit the saturating, but not the linear component of sucrose uptake in leaf tissue. PCMBS inhibited the mannitol-induced saturable component to a level equivalent to uptake in dilute buffer, but had no effect on uptake from the buffer solution (Fig. 11).

These preliminary results do not differentiate between an effect of turgor directly on the carrier or on the activity of the plasmalemma ATPase. However, since the effect appears to be specific for sucrose uptake, we conclude that the turgor effect is specifically on the sucrose carrier and not an effect on the H$^+$-ATPase.

Previous workers have shown that tissue will leak solutes at high turgor. We found the same to be true for this tissue (Table I). Tissue equilibrated in 20 milliosmolar buffer lost 19% of the tissue sucrose

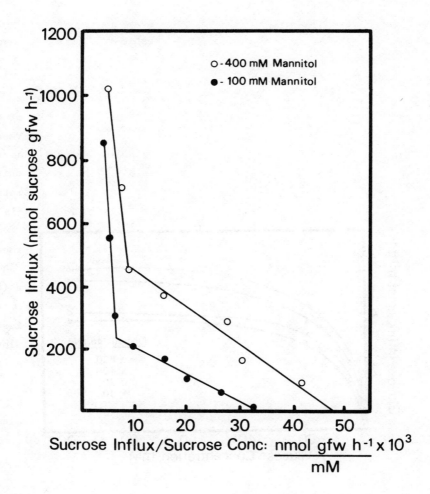

FIG. 8. Replot of kinetic profile at 100 and 400 mM mannitol using the Eadie-Hofstee transformation.

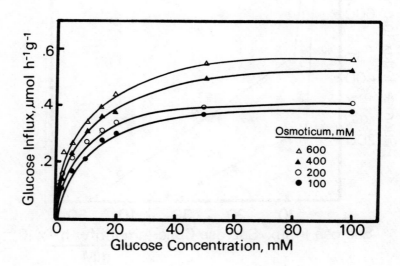

FIG. 9. Kinetic profile of [^{14}C]glucose uptake by sugar beet taproot tissue at various external osmotic concentrations. Uptake procedures were as described in Figure 4.

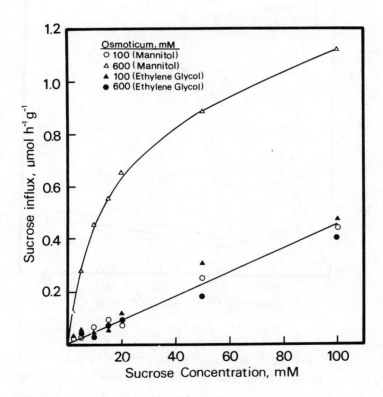

FIG. 10. Comparison of penetrating (ethylene glycol) and non-penetrating (mannitol) osmotica on the kinetics of sucrose uptake.

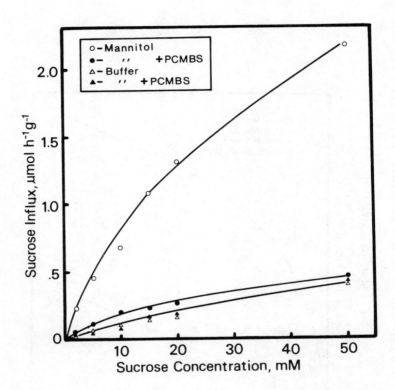

FIG. 11. Effect of PCMBS on sucrose uptake at high and low turgor. High turgor was produced in the tissue by maintenance in buffer while mannitol (500 mM) reduced the tissue turgor. The experiment was conducted as described in Figure 4, except that 1 mM PCMBS was added to the equilibration media for 30 min. The PCMBS treatment period was followed by a 30-min wash prior to incubation in [^{14}C]sucrose.

Table I. **Effect of Osmotic Concentration and PCMBS on Sucrose Efflux from Sugar beet Taproot Tissue**

Osmotic Concentration	Sucrose −PCMBS	Efflux +PCMBS
	(% of total)	
Buffer, 20 mOsm	19.0	18.6
Buffer + 500 mOsm mannitol	4.4	3.8

versus 4% when the tissue was equilibrated in 500 mM mannitol. The addition of PCMBS during the first 20 min of efflux did not inhibit sucrose loss. Therefore, the leakage is apparently not a back leak through a sucrose carrier mechanism. We were also unable to demonstrate an effect of external sucrose between 1 and 10 mM on the loss of endogenous sucrose, thus, suggesting that exchange is not an important factor in the loss of sucrose from the tissue.

Since sucrose transport across the plasma membrane is secondary active transport, the effects of turgor on the kinetics of sucrose uptake may be a result of changes in the PMF, resulting from turgor inhibition of the proton pump. To test this possibility, we ran a preliminary experiment to determine the ability of tissue to acidify the external medium over a range of cell turgors (Fig. 12). The data clearly show that at high turgor, both the extent and rate of acidification is inhibited. Thus, the changes in kinetic parameters that we observed may be due to the effect of turgor on PMF at the plasma membrane. If the PMF declines through inhibition of the ATPase at high turgor, the membrane would be unable to maintain concentration gradients and, therefore, solutes such as sucrose would leak out. The change in kinetics that we observed were due to changes in V_{max}, which could be explained by changes in the driving force of uptake, i.e. a decline in PMF.

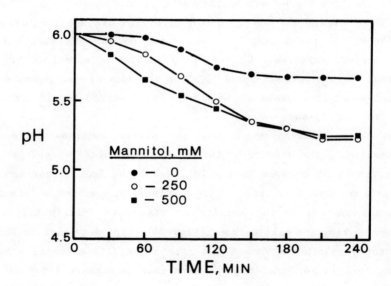

FIG. 12. Acidification of the external medium by sugar beet taproot tissue discs at high and low turgor. The tissue was equilibrated in 0.250 or 500 mM unbuffered mannitol for 60 min. At time zero, the tissue was placed in fresh medium, and the external pH monitored over time.

Plant tissues are exposed to diurnal fluctuations and water stress and, therefore, this change in turgor throughout the day may influence transport properties. In sinks which store osmotically active compounds, this effect of turgor may be important throughout a growing season.

Acknowledgments—This work was supported in part by a grant (82-CRCR-1-1074) from the USDA Competitive Research Grants Office to REW.

LITERATURE CITED

1. BENNETT AB, SD O'NEILL, RM SPANSWICK 1984 H^+-ATPase from storage tissue of Beta vulgaris. I. Identification and characterization of an anion sensitive H^+-ATPase. Plant Physiol 74:538-544
2. BENNETT AB, RM SPANSWICK 1983 Optical measurements of pH and in corn root membrane vesicles. Kinetic analysis of Cl^- effects on proton translocating ATPase. J Membrane Biol 71:95-107
3. BORCHERS-ZAMPINI C, AB GLAMM, J HODDINOTT, CA SWANSON 1980 Alterations in source-sink patterns by modifications of source strength. Plant Physiol 65:1116-1120
4. BRISKIN DP, WR THORNLEY, RE WYSE 1985 Membrane transport in isolated vesicles from sugarbeet taproot. II. Evidence of a sucrose/H^+-antiport. Plant Physiol. Submitted
5. CHURCHILL KA, B HOLAWAY, H SZE 1983 Separation of two types of electrogenic H^+-pump ATPases from oat roots. Plant Physiol 73:921-928
6. FELKER FC, JC SHANNON 1980 Movement of [^{14}C]labeled assimilates into kernels of Zea mays L. III. An anatomical examination and microautoradiographic study of assimilate transfer. Plant Physiol 65: 864-870
7. FONDY BR, DR GEIGER 1980 Effect of rapid changes in sink-source ratio on export and distribution of products of photosynthesis in leaves of Beta vulgaris L. and Phaseolus vulgaris. Plant Physiol 66:945-949
8. GIANQUINTA R 1976 Evidence for phloem loading from the apoplast. Plant Physiol 57:872-875
9. GIAQUINTA R 1977 Possible role of pH gradient and membrane ATPase in the loading of sucrose into sieve tubes. Nature 267:369-370

10. GIAQUINTA RT, W LIN, NL SADLER, VR FRANCESCHI 1983 Pathway of phloem unloading of sucrose in corn roots. Plant Physiol 72:362-367
11. GIFFORD RM, LT EVANS 1981 Photosynthesis, carbon partitioning and yield. Annu Rev Plant Physiol 32:485-509
12. KABACK HR 1983 The lac carrier protein in Escherichia coli. J Membrane Biol 76:95-112
13. KOMOR E, E ROTTER, W TANNER 1977 A proton co-transport system in a higher plant: sucrose transport in Ricinus communis. Plant Sci Lett 9:153-162
14. LEE HC, JG FORTE, D EPEL 1982 The use of fluorescent amines for the measurement of pH. Application in liposomes, gastric microsomes and sea urchin gametes. In R Nuccitelli, DW Deamer, eds, Intracellular pH: Its measurement, regulation and utilization in cellular functions. Alan Liss Inc., New York, pp 135-160
15. LEIGH RA 1983 Methods, progress and potential for the use of isolated vacuoles in studies of solute transport in higher plant cells. Physiol Plant 57:390-396
16. LICHTNER FT, RM SPANSWICK 1981 Electrogenic sucrose transport in developing soybean cotyledons. Plant Physiol 67:869-874
17. LICHTNER FT, RM SPANSWICK 1981 Sucrose uptake by developing soybean cotyledons. Plant Physiol 68:693-698
18. LIN W, MR SCHMITT, WD HITZ, RT GIAQUINTA 1984 Sugar transport into protoplasts isolated from developing soybean cotyledons. I. Protoplast isolation and general characteristics of sugar transport. Plant Physiol 75:936-940
19. MAYNARD JW, WJ LUCAS 1982 Sucrose and glucose uptake into Beta vulgaris leaf tissues. A case for general (apoplastic) retrieval systems. Plant Physiol 70:1436-1443
20. PATRICK JW 1983 Photosynthate unloading from seed coats of Phaseolus vulgaris L. General characteristics and facilitated transfer. Z Pflanzenphysiol 111:9-18
21. POOLE RJ, DR BRISKIN, Z KRATKY, RM JOHNSTONE 1984 Localization of plasma membrane and tonoplast from storage tissue of growing and dormant red beet. Characterization of proton transport and ATPase in tonoplast vesicles. Plant Physiol 74:549-556

22. SAFTNER RA, J DAIE, RE WYSE 1983 Sucrose uptake and compartmentation in sugarbeet taproot tissue. Plant Physiol 72:1-6
23. SAFTNER RA, RE WYSE 1980 Alkali cation/sucrose co-transport in the roots of sugarbeet. Plant Physiol 66:884-889
24. SMITH JC, L HALLIDY, MR TOPP 1981 The behavior of the fluorescent lifetime and polarization of oxonol potential-sensitive extrinsic probes in solution and in beef heat submitochondrial particles. J Membrane Biol 60:173-185
25. SZE H 1983 Proton-pumping adenosine triphosphatase in membrane vesicles of tobacco callus. Sensitivity to vanadate and K^+. Biochim Biophys Acta 732:586-594
26. SZE H 1984 H^+-translocating ATPases of the plasma membrane and tonoplast of plant cells. Physiol Plant 61:683-691
27. THOM M, E KOMOR 1984 H^+-sugar antiport as the mechanism of sugar uptake by sugarcane vacuoles. FEBS Lett 173:1-4
28. THOM M, E KOMOR 1984 Role of the ATPase of sugar-cane vacuoles in energization of the tonoplast. Eur J Biochem 138:93-99
29. THORNE JH 1981 Morphology and ultrastructure of maternal seed tissues of soybean in relation to the import of photosynthate. Plant Physiol 67:1016-1025
30. THORNE JH 1983 Transport of photosynthate into developing soybean seeds. Curr Top Plant Biochem 2:43-53
31. TURNER RJ 1983 Quantitative studies of co-transport systems: models and vesicles. J Membrane Biol 76:1-15
32. WYSE, R 1983 Allocation of carbon: perspective on the central role of sucrose transport and metabolism. Curr Top Plant Biochem 2:1-19

Phloem-Loading: A Metaphysical Phenomenon?

WILLIAM J. LUCAS

The term "phloem loading" was first used by Eschrich (5) to replace the then current term "vein loading." Since that time, numerous physiological and anatomical studies have been conducted with an aim to elucidating the mechanism and pathway of sugar (and amino acid) movement between the sites of production in the mesophyll cells and the sites of entry (loading) into the sieve elements. For reviews of this rather extensive literature, the reader is referred to Wardlaw (32), Geiger (11), Evert (6), Cronshaw (1), and Giaquinta (17). Our present treatment of phloem loading will not be encyclopedic in nature, but rather we will focus our attention on an evaluation of the experimental data relating to the specific point of the membrane transport mechanism involved in phloem loading.

Autoradiography still is one of the most important techniques used in the study of phloem loading/translocation. Application of either $^{14}CO_2$ or [^{14}C]-labeled sugars to mature leaf tissue indicated that the movement of radiolabeled compounds to the phloem was selective, and that the sugar, normally translocated by a particular plant, accumulated in the minor veins (8, 11). The combination of the autoradiographic image of the labeled minor veins (obtained when tissue was exposed to exogenous [^{14}C]-sucrose) and the demonstration that [^{14}C]-labeled compounds (mainly sucrose) produced by $^{14}CO_2$-mediated photosynthesis could be trapped and removed from the leaf by applying nonradioactively labeled sucrose (13), provided the basis for the hypothesis that an apoplastic stem was involved in phloem loading.

This apoplastic-step hypothesis had considerable influence on the interpretation of later work. Sovonick et al. (30) and Giaquinta (14) investigated the relationship between sugar uptake into leaf discs and the intact leaf with respect to phloem translocation. On the basis of their studies, they concluded that exogenous sucrose uptake into leaf discs reflects loading into the phloem and could, therefore, be used to study the specific mechanism involved (see also 2-4, 15, 16). Upon reflection, it is quite interesting that this concept was rarely challenged, since there were already at least two reports in the literature which showed that uptake of exogenous sucrose occurred into both the cells of the mesophyll and vascular bundles (8, 12). However, although most workers recognized that the mesophyll would act as a barrier to the penetration of exogenously supplied compounds to the phloem, it seemed that by the end of the 1970's, the general view was that phloem loading (in all tissues) could be studied via the application of $[^{14}C]$-labeled sucrose (See Fig. 1).

It was at this stage and time that our laboratory started to work on phloem physiology. Along with others working on "phloem loading," we accepted the concept of an apoplastic step and, with some slight reservations, the notion that exogenous sucrose could be used to investigate the kinetics of phloem loading, per se. In earlier studies, the kinetics of sucrose uptake were interpreted in terms of two saturable systems, both functioning to deliver sucrose into the phloem. Our initial studies, conducted on sugar beet leaf discs, simply showed that exogenous sucrose uptake occurred via two different transport systems, a saturable component displaying Michaelis-Menten kinetics and a linear component obeying first-order kinetics (26; see also 14, 21, 24). We now believe that we were incorrect in interpreting these kinetic results in terms of two transport processes operating in parallel across only the sieve element-companion cell (se-cc) complex plasma membrane (27).

RETRIEVAL OF EXOGENOUS SUGARS

It was the sugar uptake studies of Komor et al. (23), conducted on the cotyledons of germinating Ricinus seedlings, that were instrumental in causing us to re-evaluate the interpretation of our kinetic data. In

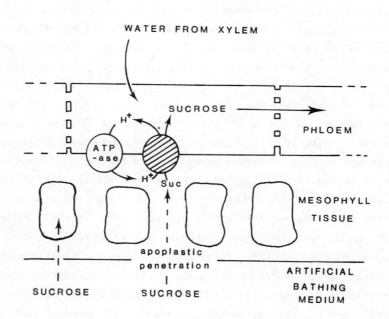

FIG 1. Schematic representation of the putative sucrose-H^+ cotransport system that has been proposed to function in phloem loading. The model implies that apoplastic or exogenously supplied [^{14}C] sucrose enters the phloem "preferentially." The broken lines depict diffusion of sucrose through the apoplast and also into the mesophyll tissue. Redrawn after Komor et al. (23).

their opinion, their linear component represented simple diffusion into cells of the mesophyll (see Fig. 1). However, they also claimed that "though it cannot be excluded that other tissues also can accumulate sucrose, phloem cells of cotyledons do it preferentially." But, in our sugar beet studies, the linear component made a significant contribution to the accumulation of sucrose by the leaf; at 25 mM exogenous sucrose it accounted for 43% of the uptake. Thus, either the linear component was not passive (since the sucrose concentration in the se-cc would be too high), or the linear component was occurring across the plasmalemma of the mesophyll cells, and, therefore, uptake could not be equated with phloem loading.

To investigate this question of location of the two transport components, we utilized leaf discs and petiole segments of sugar beet. Similarly-shaped kinetic curves were obtained for influx of sucrose, glucose, and 3-0-methyl glucose by leaf discs, whole petiole slices, petiole segments containing parenchyma tissue (pith) only, and petiole segments containing vascular bundles, although the tissues took up the various sugars via different proportions of the saturable <u>versus</u> linear uptake (see Fig. 2). These results were more consistent with the findings of Geiger (12) and Fondy and Geiger (8). They used microdensitometric techniques to examine autoradiographic plates, prepared from tissue pre-incubated for 30 min in exogenously supplied 10 mM [^{14}C] sucrose, and found that the majority of the label (60%) was in the mesophyll tissue.

The kinetic curve for 3-0-methyl glucose influx into sugar beet leaf discs was also found to be identical in form to that obtained for sucrose (see 27, Fig. 1); <u>i.e</u>. uptake appeared to result from the operation of a saturable and a linear transport component. This similarity was important because the uptake of 3-0-methyl glucose is thought to be confined to the metabolic space of the mesophyll cells (8). A similar situation was found in the petiole experiments. The petiolar vascular bundles are embedded in a parenchymatous tissue, not a chlorenchymatous mesophyll, yet sucrose uptake into vascular bundle segments still exhibited both a saturable and a linear component (see Fig. 2). Of equal importance was the finding that sucrose uptake into petiole "pith" segments (free of phloem tissues) exhibited the same kinetics as found when we used the

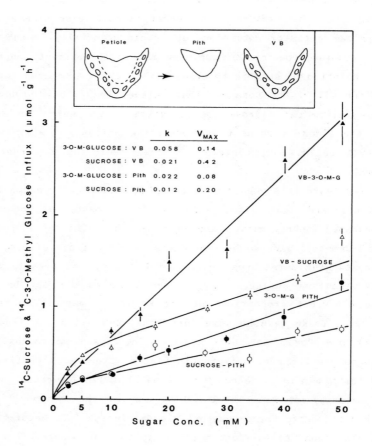

FIG. 2. Concentration dependence of sugar accumulation by dissected <u>Beta vulgaris</u> petiole segments. Points represent the mean ± SE for seven replicates. Uptake of 3-O-methyl glucose by vascular bundle-containing (VB) segments (▲); sucrose uptake by VB segments (Δ); 3-O-methyl glucose uptake by "pith" parenchyma segments (●); sucrose uptake by pith segments (O). The kinetic parameters for the saturable and linear components of the various substrates have been included; V_{max} has the units μmol g^{-1} h^{-1} and k (first order rate coefficient) has units of μmol g^{-1} h^{-1} mM^{-1} sugar. Data from Maynard and Lucas (27).

segments containing the vascular tissue. We found the same trends for 3-O-methyl glucose uptake into petiole vascular bundle--and pith--segments.

From these studies we concluded that the presence of a saturable and a linear component of transport must be a ubiquitous feature of the plasma membrane of a rather wide variety of cell types within leaf tissues. Obviously, if this were the case, it would complicate the interpretation of kinetic data with respect to the importance, etc., of a particular tissue (e.g. the se-cc complex) in terms of its contribution to the total uptake into the leaf. This, we contended, would make it very difficult to equate uptake of exogenously applied [^{14}C] sucrose to the process of phloem loading.

To resolve the question of whether all cell types of the leaf do actually possess the ability to transport exogenously applied sugars via both the saturable and linear systems, we sought a leaf from which the various tissues could be separated. We used mature leaves of Allium cepa, because it does not have an inner cuticle or epidermis, but rather it develops a lacuna; its interior tissues were, therefore, highly accessible both to exogenous solutes and cell wall digesting enzymes which were used to help to selectively remove the various cell types (Fig. 3). By filling the interior of an onion leaf with a solution of 1 mg ml^{-1} Pectolyase Y-23, we could digest away the cell walls of the large parenchyma and interveinal parenchyma cells; a 45-min digestion period was usually required. The resulting leaf, termed "pectolyase-treated" (see Fig. 3B), was then either used for [^{14}C] sucrose, or fructose kinetic studies (see Figs. 4 and 5), or further manipulated. Forceps were used to remove the vascular bundles, and the resulting leaf, termed "stripped" (see Fig. 3C), was also used in uptake experiments (33).

Figure 4 illustrates the results that we obtained for [^{14}C] sucrose influx into control and structurally manipulated leaf tissue. Clearly the curves of the inner parenchyma plus bundle sheath and the chlorenchyma have kinetic profiles that are similar to those of the vascular bundles (see inset of Fig. 4). Also, the inner parenchyma and bundle-sheath tissue accounted for ca. 70% of the total uptake, whereas the contribution of the vascular tissue was quite small (ca. 15%). Of equal importance, all the

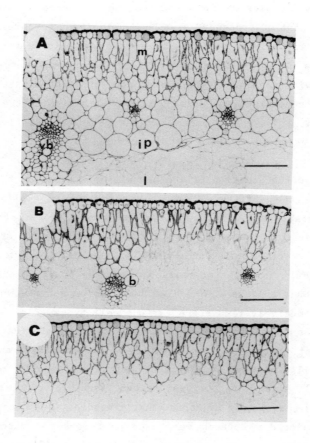

FIG. 3. Light micrographs of transverse sections taken through leaves of <u>Allium cepa</u> after they were subjected to various treatments. Control leaf (A) illustrates general arrangement of mesophyll (m), vascular bundles (vb), large inner parenchyma cells (ip), and central lacuna (l). Bar = 200 µm. In the pectolyase-treated tissue (B), the large inner parenchyma tissue is destroyed, as well as much of the bundle sheath (b). Bar = 200 µm. Forceps were used to remove the vascular bundles from pectolyase-treated tissue which then gave a leaf that was reduced to mesophyll cells only (c). Bar = 200 µm.

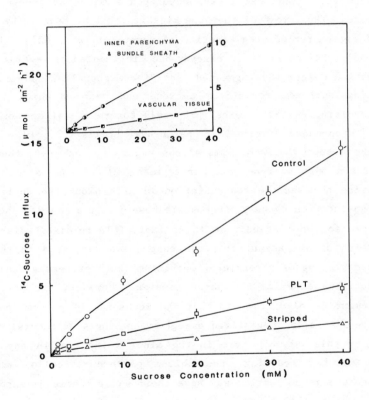

FIG. 4. Effect of pectolyase treatment (PLT) and stripping on sucrose influx into Allium cepa leaf discs. Inset: Sucrose uptake into inner parenchyma and bundle sheath cells and the vascular tissue. Data from Wilson et al. (33).

kinetic curves could be resolved into the contribution of a saturable, plus a linear (first order) component.

Similar results were obtained when we used [^{14}C] fructose, with the exception that influx into the inner parenchyma and bundle-sheath tissue contained only the linear component (see Fig. 5).

These findings provided strong support for our hypothesis that all cell types of the leaf have the ability to take up exogenously supplied sugars. We would suggest that, since sugars appear to leak out of mesophyll tissue (8), plant cells have developed a means of recovering this carbohydrate. Thus, we feel that the kinetics that we measure by applying exogenous sugars reflect the kinetics of sugar retrieval (27). In a more narrow sense, this concept had already been introduced by several workers; e.g. Fondy and Geiger (8) suggested that "the mesophyll tissue may act as a collecting surface, reclaiming sucrose and hexoses which leak from photosynthesizing cells." Other proposals for retrieval mechanisms have also been propounded for phloem parenchyma cells (7, 10, 18, 28).

If the mesophyll plasmalemma allows sugars to leak out, then we can see no reason why sucrose would not leak out of the se-cc complex. Indeed, since the sucrose concentration is high along the entire length of the translocation pathway, sucrose leakage may be a general phenomenon. Thus, at present we contend that it is impossible to distinguish between the operation of a transport process that loads sucrose into the phloem and the functioning of a retrieval mechanism; both processes would result in radioactivity entering the se-cc complex. However, the corollary to this statement is also true, in that the mechanism of phloem loading may represent a spatially-specialized aspect of an otherwise general retrieval process. By this we mean that in the minor veins, or in any location where loading is occurring, the phloem parenchyma cells immediately adjacent to the se-cc complex may have their sugar retrieval mechanism(s) repressed, while the se-cc complex may have an enhanced capacity.

Irrespective of the actual mechanism by which sugar is "loaded" into the se-cc complex of the phloem, the kinetic experiments conducted in our laboratory have shown that the in vivo kinetics of phloem loading, per se, will be extremely difficult to establish. To obtain these kinetic profiles on intact leaf tissue would require the specific inhibition of

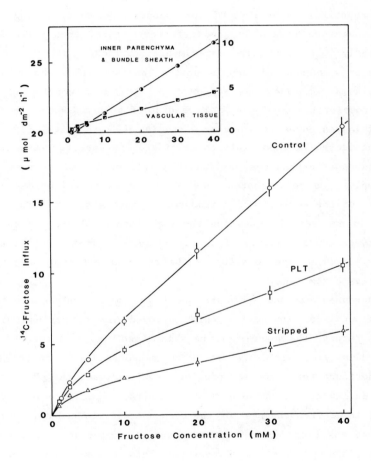

FIG. 5. Effect of pectolyase treatment (PLT) and stripping on fructose influx into Allium cepa leaf discs. Inset: Fructose uptake into inner parenchyma and bundle-sheath cells and the vascular tissue. Data from Wilson et al. (33).

all sugar retrieval (transport) systems operating into nonphloem cells. Obviously, this cannot be done at the present time and may even be impossible to achieve.

SUCROSE-H^+ CO-TRANSPORT INTO THE SE-CC COMPLEX: A RE-EVALUATION

Numerous works have been published which provide circumstantial evidence in support of a sucrose-H^+ co-transport system operating in higher plant tissues (see 17 and 29, and references cited therein). Giaquinta (17) recently evaluated the experimental evidence pertaining to the operation of a sucrose-H^+ co-transport mechanism within the phloem se-cc complex plasma membrane. He states that "the most convincing evidence for a sucrose/proton co-transport mechanism of phloem loading in photosynthesizing leaves comes from Heyser's studies on corn leaves (19)." Since we now considered it impossible to elucidate the specific kinetics of the actual transport process(es) responsible for loading of sucrose into the se-cc complex, we re-evaluated some of the experimental evidence offered in support of the sucrose-H^+ co-transport hypothesis. In this re-evaluation, we do not seek to question the experimental data, per se, nor the underlying concepts, etc., of a $\Delta\bar{\mu}_H^+$-driven system, but rather the basis for ascribing the location of the co-transport system to the se-cc complex plasma membrane.

Heyser used excised leaf strips of Zea mays, and he mounted these strips in an ingenious device which allowed him to perfuse artificial solutions through the vessels of the xylem (see Fig. 1 of ref. 19). The pH value of the xylem perfusate could be measured as it exited the leaf strip. He found that upon introducing 25 mM sucrose, the pH of the xylem solution underwent a pronounced (0.7 pH units) alkalinization, which was transient in nature; the peak of the response was reached ca. 10-min after the sucrose was introduced. Other sugars like glucose, fructose, raffinose, etc., did not elicit a pH response. This sucrose-induced pH response is certainly consistent with the operation of a sucrose-H^+ co-transport system (31), but Heyser's results do not reveal the actual location of this transport process.

This question of location was addressed by Fritz et al. (10), who used exactly the same experimental material and system as used originally by Heyser. By using the Fritz microautoradiography technique (see ref. 9), they were able to demonstrate that after an 8- to 15-min exposure to [^{14}C]-labeled sucrose (fed through the xylem perfusate), the majority of the radioactivity was located in the parenchyma cells neighboring the xylem vessels (see Fig. 6A and B). Even after a 30-min exposure, little of the label in the parenchyma cells moved into the thick- or thin-walled sieve tubes (E. Fritz, personal communication). For a comparison, we have included autoradiographs of Zea mays that were prepared from tissue that was fed $^{14}CO_2$ for 5 min, with a 10-min chase, under two different light regimes of 100 (Fig. 6C) and 250 μmol photons m^{-2} s^{-1} (Fig. 6D). These micrographs illustrate that when sugars are produced via photosynthesis, the radioactivity enters preferentially into the sieve tubes, and this is especially true when the leaf has been exposed to low light conditions (cf. Fig. 6, C and D).

Thus, the physiological evidence which Giaquinta (17) cited as being the most convincing in support of sucrose-H^+ co-transport into the se-cc complex is probably not occurring at this location, but rather at the plasmalemma of the xylem (or vascular) parenchyma. We agree with the interpretation offered by Fritz et al. (10) that these cells probably function to retrieve sucrose from the xylem vessels.

A second plant tissue that has been used to study the operation of the putative sucrose-H^+ co-transport system in the phloem is the cotyledons of germinating Ricinus seedlings (20, 22, 23). The reader will perhaps recall that Figure 1 is an illustration of what Komor and his coworkers presented for sucrose uptake into this tissue (23). Martin and Komor (25) and Komor et al. (23) used autoradiography of entire cotyledons to correlate the presence of [^{14}C] sucrose with the sucrose-induced alkalinization of the experimental medium (see also Ref. 20). Although Martin and Komor (25) reported that after 5- or 10-min exposure periods the radioactive label appeared to be distributed over the entire cotyledon, they suggested that this was probably an artifact in that the minor veins could not be detected, rather than being due to label in the mesophyll cells. These authors did say that a role of the mesophyll in sucrose uptake could not be excluded, although they thought it very unlikely. However, Komor et al. (23) stated that since "phloem cells of cotyledons

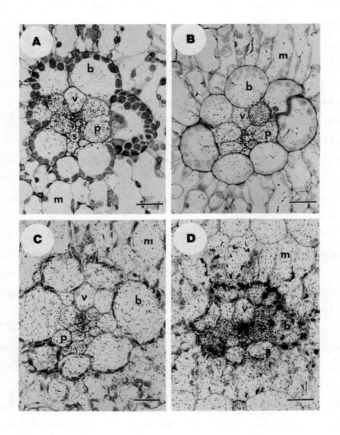

FIG. 6. Microautoradiographs of 1-µm transverse sections of Zea mays leaf strips; b, bundle-sheath cells; m, mesophyll cell; p, vascular parenchyma cell; s, thin-walled sieve tube; v, xylem vessel; all bars = 20 µm. A: leaf strip fed 2.5 mM [^{14}C] sucrose via the xylem (using the technique of Heyser [19]); feeding time, 8 min. B: As in A, except feeding time was 15 min. C and D: Leaf strips exposed to $^{14}CO_2$ for 5-min followed by a 10-min pulse with $^{12}CO_2$. Two light intensities were employed; C, 100 µmol photons m^{-2} s^{-1}; D, 250 µmol photons m^{-2} s^{-1}. (Light micrographs kindly provided by Dr. E. Fritz, der Universität Göttingen.)

do it preferentially," the "most simple assumption would be that it is <u>direct</u> uptake into the phloem cells which proceeds by proton symport."

In a collaborative study that is presently being conducted by Dr. Eberhardt Fritz and Dr. Ewald Komor, these workers are now applying the Fritz microautoradiographic technique to the <u>Ricinus</u> cotyledon. As with the work on <u>Zea mays</u>, Dr. Fritz is examining the distribution of the radioactivity in exactly the same system as previously used by Komor and his co-workers. The results so far obtained are extremely interesting. In tissue exposed to [^{14}C]-labeled sucrose for up to 30-min, <u>i.e.</u> the same time as needed to reach the peak of the alkalinization response, the majority of the label has been found in the mesophyll tissue (E. Fritz, personal communication). These microautoradiographic results would tend to suggest that the sucrose-H^+ co-transport system functions across the plasma membrane of the mesophyll cells. Thus, the contribution of the phloem cells to the measured pH transient may be quite small. Indeed, it may be impossible to detect the contribution of this tissue due to the rather long diffusion path from the se-cc complex plasma membrane to the external bathing medium.

ELECTROPHYSIOLOGICAL STUDIES ON PHLOEM CELLS

Perhaps one of the most interesting studies on the physiology of phloem cells was that conducted by Wright and Fisher (34). These workers developed an elegant technique that allowed them to make electrophysiological measurements on phloem tissue. To achieve this, they employed aphids that probed the phloem of willow stems (bark); once the aphid had penetrated the sieve element, it was anaesthetized and in this state its stylet was cut. The cut stylet was then used to gain electrical continuity with the inner surface of the sieve element plasma membrane.

The values of the membrane potential that were measured with this technique were within the range of values normally measured on higher plant tissues. More importantly, Wright and Fisher (34) were able to show that when exogenous sucrose was introduced into their system, a transient depolarization of the membrane potential was observed; the extent of the depolarization was a function of the concentration of sucrose present in the bathing medium. These results were interpreted as providing direct evidence for the operation of a sucrose-H^+ co-transport mechanism in the plasma membrane of the willow sieve element. However, there is a problem with this interpretation. Application of other sugars, like fructose, 3-0-methyl glucose and glucose, also elicited the same type of transient depolarization in the measured membrane potential. The extent of the depolarization was ca. 50 to 70% of that obtained when sucrose was employed.

Earlier autoradiographic studies on leaf tissue have shown that [^{14}C]glucose and ^{14}C-3-0-methyl glucose do not enter the phloem when they are applied to the apoplast, but rather are taken up by the mesophyll cells (8). If this were also true for the willow bark system employed by Wright and Fisher, their electrical response, obtained with these non-translocatable sugars, may be located at the plasma membrane of the surrounding vascular parenchyma. Plasmodesmatal connections between the se-cc complex and the vascular parenchyma would need to exist to permit the transmission of this electrical signal from the parenchyma to the electrode inserted into the aphid stylet. If this were the case for glucose, 3-0-methyl glucose and fructose, it would make it very difficult to argue that the same situation does not also exist with respect to the sucrose-induced transients. Again, the results are consistent with the operation of a plasma membrane-bound sugar-H^+ co-transport mechanism, within the tissue of the phloem, but we consider that it does not provide direct evidence for the presence of such a system within the sieve element plasma membrane.

THE QUESTION OF PHLOEM LOADING

In view of these recent findings, we are forced to ask ourselves just what do we really know about the phenomenon of phloem loading. We know that certain sugars do accumulate within the se-cc complex and the extent of this accumulation indicates that an active process seems to be required. We contend that the exact nature of this active process is yet to be elucidated. We agree that since many plant tissues (mesophyll, vascular parenchyma, scutellum, etc.) appear to contain a sucrose-H^+ co-transport mechanism, such a system may also function at the se-cc complex plasma membrane. But, at the present time, we cannot provide direct evidence in support of this hypothesis. This is why we have raised the point that phloem loading may be a "metaphysical phenomenon."

The solution to this dilemma may lie in the utilization of a much wider range of plant systems than is presently being employed in studies on "phloem loading." It may be possible to utilize the biochemical variability present in the various leaf types, in combination with genetic, physiological and environmental manipulations, to establish a system that will allow us to conduct direct studies on the phloem. Whether or not this will be possible will be illustrated by future experiments.

Acknowledgments--The work conducted in our laboratory on phloem physiology has been made possible by National Science Foundation grants PCM 79-10223 and PCM 83-15408. Dr. Eberhardt Fritz generously made available his published and unpublished microautoradiographic results on Zea mays and Ricinus. Stimulating discussion with all my sabbatical colleagues in the Forstbotanisches Institut, der Universität Göttingen is gratefully acknowledged. Finally, I would like to thank Dr. John Oross for his help with the micrographs and Lyn Noah for typing the manuscript.

LITERATURE CITED

1. CRONSHAW J 1981 Phloem structure and function. Annu Rev Plant Physiol 32:465-484
2. DELROT S 1981 Proton fluxes associated with sugar uptake in Vicia faba leaf tissue. Plant Physiol 68:706-711
3. DELROT S, JL BONNEMAIN 1981 Involvement of protons as a substrate for the sucrose carrier during phloem loading in Vicia faba leaves. Plant Physiol 67:560-564
4. DELROT S, JP DESPEGHEL, JL BONNEMAIN 1980 Phloem loading in Vicia faba leaves: effect of N-ethylmaleimide and parachloromercuribenzene sulfonic acid on H^+ extrusion, K^+ and sucrose uptake. Planta 149:144-148
5. ESCHRICH W 1970 Biochemistry and fine structure of phloem in relation to transport. Annu Rev Plant Physiol 21:193-214

6. EVERT RF 1977 Phloem structure and histochemistry. Annu Rev Plant Physiol 28:199-222
7. EVERT RF, W ESCHRICH, W HEYSER 1978 Leaf structure in relation to solute transport and phloem loading in Zea mays L. Planta 138: 279-294
8. FONDY BR, DR GEIGER 1977 Sugar selectivity and other characteristics of phloem loading in Beta vulgaris L. Plant Physiol 59:953-960
9. FRITZ E 1980 Microautoradiographic localization of assimilates in phloem: problems and new method. Ber Dtsch Bot Ges 93:109-121
10. FRITZ E, RF EVERT, W HEYSER 1983 Microautoradiographic studies of phloem loading and transport in the leaf of Zea mays L. Planta 159:193-206
11. GEIGER DR 1975 Phloem loading. In MH Zimmerman, JA Milburn, eds, Transport in Plants, Encyclopedia of Plant Physiology, Vol 1, New Series. Springer-Verlag, Heidelberg, pp 395-431
12. GEIGER DR 1976 Phloem loading in source leaves. In IF Wardlaw, JB Passioura, eds, Transport and Transfer Processes in Plants. Academic Press, New York, San Francisco, London, pp 167-215
13. GEIGER DR, SA SOVONICK, TL SHOCK, RJ FELLOWS 1974 Role of free space in translocation in sugar beet. Plant Physiol 54:892-899
14. GIAQUINTA RT 1977 Phloem loading of sucrose: pH dependence and selectivity. Plant Physiol 59:750-755
15. GIAQUINTA RT 1979 Phloem loading of sucrose: involvement of membrane ATPase and proton transport. Plant Physiol 63:744-748
16. GIAQUINTA RT 1980 Mechanism and control of phloem loading of sucrose. Der Dtsch Bot Ges 93:187-201
17. GIAQUINTA RT 1983 Phloem loading of sucrose. Annu Rev Plant Physiol 34:347-387
18. GILDER J, J CRONSHAW 1974 A biochemical and cytochemical study of adenosine triphosphatase activity in the phloem of Nicotiana tabacum. J Cell Biol 60:221-225
19. HEYSER W 1980 Phloem loading in the maize leaf. Der Dtsch Bot Ges 93:221-228

20. HUTCHINGS VM 1978 Sucrose and proton cotransport in Ricinus cotyledons. I. H^+ influx associated with sucrose uptake. Planta 138: 229-235
21. KOMOR E 1977 Sucrose uptake by cotyledons of Ricinus communis L.: characteristics, mechanism, and regulation. Planta 137:119-131
22. KOMOR E, M ROTTER, W TANNER 1977 A proton-cotransport system in a higher plant: sucrose transport in Ricinus communis. Plant Sci Lett 9:153-162
23. KOMOR E, M ROTTER, J WALDHAUSER, E MARTIN, BH CHO 1980 Sucrose proton symport for phloem loading in the Ricinus seedling. Der Dtsch Bot Ges 93:211-219
24. LICHTNER FT, RM SPANSWICK 1981 Sucrose uptake by developing soybean cotyledons. Plant Physiol 68:693-698
25. MARTIN E, E KOMOR 1980 Role of phloem in sucrose transport by Ricinus cotyledons. Planta 148:367-373
26. MAYNARD JW, WJ LUCAS 1982 A reanalysis of the two-component phloem loading system in Beta vulgaris. Plant Physiol 69:734-739
27. MAYNARD JW, WJ LUCAS 1982 Sucrose and glucose uptake into Beta vulgaris leaf tissues: A case for general (apoplastic) retrieval systems. Plant Physiol 70:1436-1443
28. PATE JS, BES GUNNING 1969 Vascular transfer cells in angiosperm leaves. A taxonomic and morphological survey. Protoplasma 68: 135-156
29. REINHOLD L, A KAPLAN 1984 Membrane transport of sugars and amino acids. Annu Rev Plant Physiol 35:45-83
30. SOVONICK SA, DR GEIGER, RJ FELLOWS 1974 Evidence for active phloem loading in the minor veins of sugar beet. Plant Physiol 54:886-891
31. TANNER W 1980 Proton sugar cotransport in lower and higher plants. Ber Dtsch Bot Ges 93:167-176
32. WARDLAW IF 1974 Phloem transport: physical, chemical or impossible. Annu Rev Plant Physiol 25:515-539
33. WILSON CW, JW OROSS, WJ LUCAS 1985 Sugar uptake into Allium cepa leaf tissue: An integrated approach. Planta. In press
34. WRIGHT JP, DB FISHER 1981 Measurement of the sieve tube membrane pontential. Plant Physiol 67:845-848

Carroll A. Swanson

We honor Carroll Swanson in this year of his retirement for his many y(ears) of leadership in translocation research. Through much of his professi(onal) life, translocation studies were considered by many to be relatively unim(por)tant. Yet, Carroll continued his efforts, for he was convinced of the imp(ort)ance of his work. We believe that this Symposium is a confirmation of (his) conviction.

Carroll has influenced many of the participants in this Symposium. Sev(eral) of us worked directly with him as students at The Ohio State University. (Not) only did we learn about translocation, but we were influenced by his do(gged) insistance on careful, thorough research. He also taught us to communicate (our) research results clearly. Many others have been influenced indirectly by wo(rk)ing with Carroll's students. In addition, all who have done research on tr(ans)location or taught the subject have depended on his outstanding publicati(ons).

Though Carroll was not a direct contributor to this Symposium, he did c(on)tribute through many of us. As children honor their genetic parents (for) their biological lives, we honor Carroll Swanson for our professional li(ves).

> John E. Hendrix
> Colorado State University

> Donald R. Geiger
> University of Dayton

Partitioning of Sucrose into Fructans in Winter Wheat Stems

JOHN E. HENDRIX

There are a number of steps or processes that govern source-sink relationships. Several of these processes have already been discussed in detail during this symposium. The one I wish to emphasize is the ability of various sink processes to compete for photoassimilate. An important component of that competition is the ability of sink cells to sequester translocated carbons into locations or substances that make the carbons unavailable to the translocation system. Those processes maintain a free energy gradient for the translocated materials from sieve tubes into sink cells. The most obvious systems are those that produce starch in sink cells. Other systems, such as the sugar beet root, transfer sucrose into vacuoles. However, systems that stored osmotically active materials generate larger turgor in sink cells. As Wyse points out (see this symposium, p. 249), the ability of sink cells to absorb translocated material appears to be suppressed by high turgor. Therefore, the ability of sink cells to convert sucrose into substances of lower osmotic activity per carbon should be related to a greater sink capacity and/or strength.

There are both applied and basic implications to our being able to understand and control these partitioning systems. Examples of genetic control of these systems was demonstrated by Austin et al. (1) and further developed by Gifford et al. (6). They pointed out that plant breeders have serendipitously selected for altered partitioning in their efforts to select for greater economic productivity. One major step in this progression was the development of semi-dwarf grain cultivars which partition less carbon into structural components of the stem.

UTILIZATION OF TRANSLOCATED SUCROSE

The work I will report here focuses on the synthesis of nonstructural carbohydrates (NSC) in winter wheat stems. The initial effort was directed toward understanding the utilization of translocated sucrose by developing wheat grains on intact plants. We needed to learn whether we could introduce asymmetrically labeled sucrose into flag leaves and have that sucrose translocated in an unaltered form. We succeeded in doing that. A minor part of that effort involved supplying [^{14}C]glucose to the plants through flag leaves 10 d post-anthesis, a time of rapid grain growth. As expected, the label from the glucose became randomized between the two moieties of sucrose, at least near the radioactive front (13) (see Fig. 1). However, nearer the supply point, the sucrose was more heavily labeled in its glucose moiety. Analogous data, though more difficult to interpret, were obtained when [^{14}C](glucosyl)sucrose was supplied (12). We concluded that the stem was acting as a sink; and, in addition, that the fructose moiety of sucrose was being utilized selectively. We believe that new sucrose was being synthesized using the residual [^{14}C]glucose and fructose from a pool that lagged in its approach to isotropic equilibrium (12). An appropriate sink mechanism for the selective use of the fructose moiety of sucrose would be fructan synthesis as indicated below:

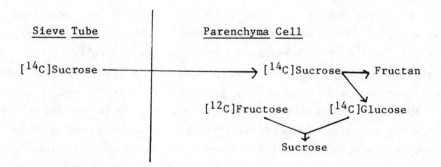

As indicated by Smith (18) and Meier and Reid (11), fructans are the most abundant NSC in the stems of wheat, oats, barley, rye, and cool-season perennial grasses. Starch is generally reported to be absent or in

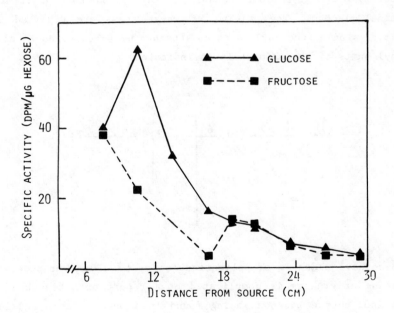

FIG. 1. ^{14}C in hexose moieties of sucrose as a function of distance from the labeling source and supply on flag leaf of winter wheat. ^{14}C supplied as [^{14}C]glucose 80 min before harvest.

limited quantity. However, we are aware of one report that wheat stems contain significant quantities of starch (8). We investigated that possibility on samples of four wheat cultivars harvested at anthesis. After stems were homogenized in 90% ethanol (v/v), no microscopically detectable indication of starch was obtained after treatment of the residue with I_2/KI solution. Yet, the stems were green at the time of harvest, indicating the presence of chloroplasts.

FRUCTAN BIOCHEMISTRY

According to Meier and Reid (11), fructans are of several configurations. The most studied is inulin, produced by a number of species, including Jerusalem artichoke. Inulin is an unbranched polymer of fructose connected by β-2,1-glycosidic bonds with a terminal or subterminal glucose. The fructans of grasses are phleins, with their fructose moieties connected by β-6,2-glycosidic bonds. Phleins also have a terminal or subterminal glucose. The phleins of wheat stems are reported to have branches of single fructosyl moieties attached by β-1,2-bonds to alternate fructosyl units of the main chain as indicated:

$$\text{Glu} \xrightarrow{1 \quad 2} \text{Fru} \xrightarrow{6 \quad 2} \left[\text{Fru} \xrightarrow{6 \quad 2} \text{Fru} \atop {\begin{array}{c} 1 \\ 2 \\ \text{Fru} \end{array}} \right]_n$$

These have a degree of polymerization up to 250. Our understanding of fructans of grasses is complicated by a report of a branched fructan with an inulin-type backbone being found in a wheat flour, and in stems of barley and rye (16).

Enzymology for inulin synthesis by two enzymes of Jerusalem artichoke was proposed by Edelman and co-workers (3, 4), as indicated below:

Enzymes Proposed for Fructan Synthesis

Sucrose-sucrose fructosyl transferase [I]

$$2 \text{ Sucrose} \longrightarrow \text{G-F-F} + \text{G} \qquad (1)$$

β-(2-1) fructan: β-(2-1) fructan 1-fructosyl transferase [II]

$$\text{G-F-F} + \text{G-F-F} \longrightarrow \text{G-F}$$
$$+ \text{G-F-F-F} \qquad (2)$$
$$\text{G-F-F-F-F} + \text{G-F}$$
$$\longrightarrow \text{etc.}$$

The first, sucrose-sucrose fructosyl transferase (SSTase) transfers a fructose moiety from one sucrose molecule to another, forming a trisaccharide and glucose. The second enzyme, β-(2-1) fructan: β-(2-1) fructan 1-fructosyl transferase (FFTase) transfers a fructosyl from one trisaccharide to another, forming a tetrasaccharide and sucrose (11). It was proposed that this mechanism of FFTase continues to catalyze polymer elongation. Wagner et al. (19) recently isolated SSTase from wheat leaves.

LABELING FRUCTANS WITH TRANSLOCATED SUCROSE

Even though SSTase and FFTase do not account for the formation of branched fructans, we proposed that a mechanism, similar to that indicated above, operates in wheat stems. To test that hypothesis, the flag leaves of wheat plants were supplied with a mixture of (glucose-^{14}C(U))-sucrose and (fructose-1-^{3}H)-sucrose at anthesis. Three hours later, stems were cut into three sections per internode. The stem pieces were extracted with 80% ethanol (v/v), then with water. The water extracts were dried, then hydrolyzed with trifluoroacetic acid. The resultant hexoses were separated by paper chromatography, then counted. Representative data from one plant are presented graphically in Figures 2 and 3, and tabular data from additional plants are presented in Table I.

Figure 2 represents the data from the hydrolyzed water extract of the top section of the second internode (counting down from the head). The ^{3}H:^{14}C ratio of sucrose supplied to that plant was 6.00. Note that very little ^{14}C was incorporated into the fructose moieties (ratio of ^{3}H:^{14}C was 37) and little ^{3}H into the glucose portion (ratio was 0.37), indicating that little randomization could have occurred and that sucrose was transferred to sink cells unaltered.

The data of Figure 3 were derived from the lower third of the upper internode. Those tissues were immature. There was significant randomization of label in that section (note fructose ^{3}H:^{14}C ratio). We hypothesize metabolism in immature tissues was directed more toward growth than storage, so translocated sucrose was degraded soon after or even before entering parenchyma cells. What sucrose was available for fructan synthesis resulted from resynthesis after the label had been randomized. Obviously, as indicated by the ^{3}H:^{14}C ratio for glucose, randomization was not complete. Data (not shown) for other stem sections of that plant (but see Ref. 7) and data from other plants (Table I) are consistent with the above explanation. It would be interesting to compare activity of cell wall invertase from mature and immature tissue.

For the more mature stem sections, we propose the partitioning of carbons from translocated sucrose, as indicated in Figure 4. This is similar to the systems of Kandler and Hopf (9), as modified from Edelman and Jefford (4). As indicated, we believe that phlein synthesis in wheat stems is similar to the mechanism proposed for inulin synthesis, in that both require the direct transfer of the fructose moiety of sucrose into the growing polymer. However, this does not explain the branched structure proposed for wheat stem phleins.

Robyt and Martin (14) found a similarly branched glucan in <u>Streptococcus mutans</u> 6715. Sucrose is the substrate for synthesis of this glucan and a mechanism for synthesis has been proposed by Robyt and co-workers (2, 15). The mechanism of phlein synthesis may be similar.

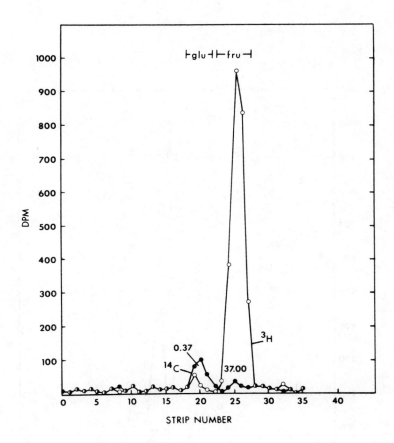

FIG. 2. Chromatograms of radioactivity found in hydrolyzed water extract from top third of second internode (counting from head) of wheat plant. DPM was measured per 1-cm strips of the chromatogram. Material harvested 3 h after mixture of (glucose-^{14}C(U))-sucrose and (fructose-1-^{3}H)-surcose was supplied to flag leaf. Supplied ^{3}H:^{14}C ratio was 6.00. Ratio within peaks as indicated.

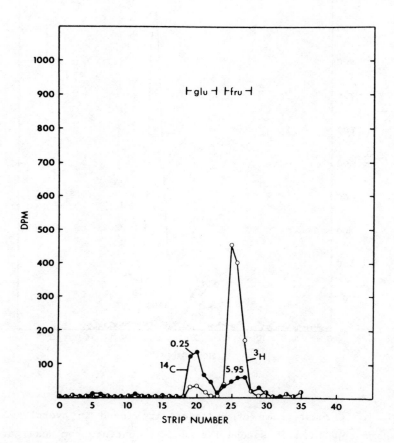

FIG. 3. Chromatograms of radioactivity found in hydrolyzed water extracts of lower third of top internode. Other parameters are as described in Figure 2.

Table I. Ratios of ^{3}H:^{14}C in Supplied Sucrose and Hexoses from Hydrolyzed Water Extracts of Stem Tissue

Experiment/ Sample	^{3}H:^{14}C Supplied	^{3}H:^{14}C- Glucose	^{3}H:^{14}C- Fructose
Plant 2[a]			
1-3		0.38	5.02
	5.04		
2-1		0.56	20.63
Plant 3			
1-3		0.72	2.60
	2.67		
2-1		0.21	44.50
Plant 4			
1-3		0.49	3.80
	3.85		
2-1		0.87	18.92
Plant 5			
1-3		0.59	3.63
	3.62		
2-1		0.19	36.55
Plant 6			
1-3		0.42	4.04
	4.05		
2-1		0.37	27.31

[a] Stem sections 1-3 and 2-1 (above and below the node of insertion of the flag leaf, respectively) are listed for each plant.

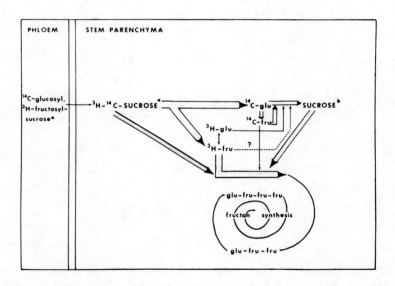

FIG. 4. Proposed pathway for hexose moiety of sucrose into fructans of wheat stems. SUCROSE[a], asymmetrically labeled sucrose as supplied to plant. SUCROSE[b], sucrose resynthesized from supplied hexose moieties.

COMPETITION BETWEEN STEMS AND INFLORESCENCES FOR TRANSLOCATED SUCROSE

How is this relevant to sucrose partitioning? The common view has been that more NSC accumulated in stems prior to grain filling results in greater yield. The basis of this idea is that these NSC are mobilized during grain filling, thereby contributing to grain yield (20). However, Fischer (5), using spring wheat, demonstrated that limiting photoassimilate by shading inhibited grain production most severely if the shading occurred during floral development. In addition, Shanahan et al. (17), using winter wheat, have shown that grain number is more strongly correlated with grain yield than is grain size (Table II). These data indicate that the capacity of the grains as a sink is controlled primarily by grain number. Furthermore, Koomanoff (10) has presented data indicating a highly significant negative correlation between grain number and stem NSC 6 d post-anthesis.

It is during floral development that the maximum potential for grain number is established. Also, during that time, there is rapid stem elongation as well as NSC accumulation in stems. All of the above-defined correlations are justification for studying the synthesis of fructans, the prime NSC of wheat stems.

Using the International Wheat Yield Nursery at Colorado State University, last spring (1984), we harvested material from four cultivars. Material was harvested every week, starting at the time stem elongation was initiated and extending to anthesis. We also harvested grain at maturity. Heads and stems were ground separately in 90% ethanol (v/v) and filtered. The residue was extracted with water at 60 to 65°C. We have analyzed most of the water extract using a fructose-specific anthrone method (analysis at room temperature).

Data from those analyses are presented in Figure 5 as per cent fructan (dry weight) compared to grains per head. Data from two harvest dates are indicated.

We believe that fructan synthesis prior to anthesis competes with the developing inflorescence for translocated sucrose. Analysis at anthesis supplies data on the net fructan synthesis that occurred during that critical floral development period. However, it should be emphasized that these data are limited and preliminary.

Table II. Linear Correlation Coefficients for Grain Yield Versus Grain Number and Grain Yield Versus Grain Size for Winter Wheat

Location and Year	Kernel No. Versus Grain Yield	Kernel Size Versus Grain Yield
Adams Co., CO (1977)†	0.862**	-0.159
Sedgwick Co., CO (1977)	0.813*	0.182
Adams Co., CO (1978)	0.974**	0.027
Wash. Co., CO (1978)	0.112	0.686*
Logan Co., CO (1978)	0.391	-0.013
Sedgwick Co., CO (1978)	0.782**	-0.273
Across all Colorado sites	0.802*	-0.196
Weslaco, TX (1978)	0.881**	0.771**
Weslaco, TX (1979)	0.873**	0.304
Bushland, TX (1979)	0.772*	0.050
Pendleton, OR (1974)	0.889**	0.583

*, ** Significant at the 5 and 1% levels, respectively.

† Colorado sites from original data of Shanahan et al. (17), other sites data from literature developed by Shanahan et al. (see their article for citations).

In computing the regression lines, the data for Katya (d) at anthesis were not used, and the data for Colt (c) were not used at pre-anthesis. This exclusion of the Colt data at pre-anthesis is based on its slower rate of maturity. The pre-anthesis harvest was 30 May. The anthesis harvest dates for all cultivars except Colt were 6 or 7 June. The anthesis date for Colt was 11 June. The fructan concentration for all varieties was below 3% stem dry weight at the 23 May harvest, indicating that most of the fructan synthesis occurred in the last 10 to 14 d prior to anthesis. For Colt, the 30 May harvest was 12 d before its anthesis date, likely about the time rapid fructan synthesis was starting, while other

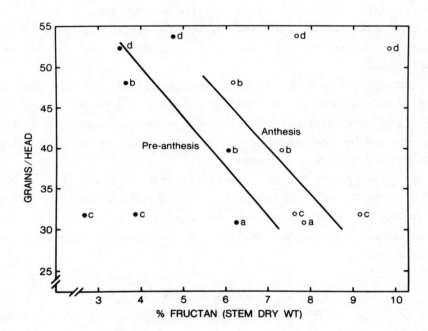

FIG. 5. Comparison of grains per head to fructan concentration. Open circles, anthesis; closed circles, pre-anthesis (see Text for dates). Points c not used for pre-anthesis regression line, points d not used for anthesis regression line. Cultivar replicates indicated a, TAM 107; b, Dobroudja; c, Colt; d, Katya A-1. Correlation coefficients are −0.82 and −0.81 for anthesis and pre-anthesis, respectively (both significant at 0.10).

varieties were well into their fructan synthesizing phase by the 30 May harvest. If one makes the simple assumption that the increase of stem fructan content was linear in Colt (c) from 30 May to 11 June, the concentration of fructans 7 d pre-anthesis was 5.3% and 5.4% for the replicate samples. Those values are in reasonable agreement with the values for the other cultivars.

The exclusion of the data from Katya (d) at anthesis is more arbitrary. However, Katya is morphologically quite distinct from the other cultivars. It may be so distinct genetically that it should be compared with a different set of cultivars. However, it should not be eliminated from further work, for it may have systems for photoassimilate production and partitioning that are sufficiently distinct to warrant detailed study.

Although our research is still preliminary, we feel that an understanding of the control of this system will contribute to our understanding of partitioning. We will need to study the enzymology of fructan synthesis and degradation to be able to understand how the plant maintains gradients for sucrose across membranes.

Acknowledgments—For participation in various parts of this work, I wish to thank N. Robinson (UC Davis), M. E. Hogan, D. Smith, C. Ross, J. Linden, and J. Quick. This work was supported in part by the Colorado State University Experiment Station and published as Scientific Series Paper No. 2981.

LITERATURE CITED

1. AUSTIN R, J BINGHAM, RD BLACKWELL, LT EVANS, MA FORD, CL MORGAN, M TAYLOR 1980 Genetic improvement in winter wheat yields since 1900 associated with physiological changes. J Agric Sci 94:675-689
2. COTE CL, JF ROBYT 1982 Acceptor reaction of alteransucrose from Leuconostoc mesenteroides NRRL B-1355. Carbohydr Res 111:127-142
3. EDELMAN J, AG DICKERSON 1966 The metabolism of fructose polymer in plants. Transfructosylation in tubers of Helianthus tuberosus L. Biochem J 98:787-794

4. EDELMAN J, TG JEFFORD 1968 The mechanism of fructosan metabolism in higher plants as exemplified by Helianthus tuberosus L. New Phytol 67:517-531
5. FISCHER FA 1975 Yield potential in a dwarf spring wheat and the effect of shading. Crop Sci 15:607-613
6. GIFFORD RM, JH THORNE, WD HITZ, RT GIAQUINTA 1984 Crop productivity and photoassimilate partitioning. Science 225:801-808
7. HOGAN ME 1983 Labeling of nonstructural carbohydrates in winter wheat stems. M.S. thesis, Colorado State University, Fort Collins, CO
8. JUDELL GK, K MENGEL 1982 Effect of shading on nonstructural carbohydrates and their turnover in culms and leaves during the grain filling period of spring wheat. Crop Sci 22:958-962
9. KANDLER O, H HOPF 1982 Oligosaccharides based on sucrose (sucrosyl oligosaccharides). In F Loewus, W Tanner, eds, Plant Carbohydrates. I Intracellular Carbohydrates, Encyclopedia of Plant Physiology, Vol 13A, New Series. Springer Verlag, Berlin, pp 348-383
10. KOOMANOFF NE 1981 Relationship among stem carbohydrate reserves grain yield and yield components in winter wheat. M.S. thesis, Colorado State University, Fort Collins, CO
11. MEIER H, JSG REID 1983 Reserve polysaccharides other than starch. In F Loewus, W Tanner, eds, Plant Carbohydrates. I Intracellular Carbohydrates, Encyclopedia of Plant Physiology, Vol 13A, New Series. Springer Verlag, Berlin, pp 418-476
12. ROBINSON NL 1980 Translocation of ^{14}C-sucrose in wheat. M.S. thesis, Colorado State Univ, Fort Collins, CO
13. ROBINSON NL, JE HENDRIX 1983 Translocation of [^{14}C]sucrose in wheat. Plant Physiol 71:701-702
14. ROBYT JF, PJ MARTIN 1983 Mechanism of synthesis of D-glucans by D-glucosyltransferase from Stroptococcus mutans 6715. Carbohydr Res 113:301-315
15. ROBYT JF, TF WALSETH 1978 The mechanism of acceptor reactions of Leuconostoc mesenteroides B-512 F dextransucrose. Carbohydr Res 61:433-435

16. SCHLUBACH HH, F LEDERER 1960 Untersuchungen uber polyfructosane LVIII. Der Kohlenhydrastoffwechsel in Weisen. Ann Chem 635:154-165
17. SHANAHAN JF, DH SMITH, JR WELSH 1984 An analysis of post-anthesis sink-limited winter wheat grain yields under various environments. Agron J 76:611-615
18. SMITH D 1973 The nonstructural carbohydrates. In GW Beutler, RW Bailey, eds, Chemistry and Biochemistry of Herbage, Vol 1. Academic Press, NY, pp 105-166
19. WAGNER W, F KELLER, A WIEMKIN 1983 Fructan metabolism in cereals: induction in leaves and compartmentation in protoplasts and vacuoles. Z Pfanzenphysiol 112:359-372
20. YOSHIDA S 1972 Physiological aspects of grain yield. Annu Rev Plant Physiol 23:437-464

Significance of Carbon Allocation to Starch in Growth of *Beta vulgaris* L.

DONALD R. GEIGER, LEANNE M. JABLONSKI, and BERNARD J. PLOEGER

Several recent studies provide evidence that the allocation of newly-fixed carbon to starch in source leaves may be an important factor determining root:shoot ratio (6, 9). In that case, partitioning of carbon between root and shoot depends not only on processes that partition carbon among sink regions, but also on the distribution of newly-fixed carbon between storage carbon for delayed export and that available for immediate export. For example, when shifted directly from long to short days, plants of several species, including sugar beet and soybean, show a several-fold increase in the rate of leaf-starch accumulation (1-4). If the proportion of carbon partitioned between root and shoot differs between day and night, as recent data suggest (5, 6, 9), a change in the rate of allocation of newly-fixed carbon to starch could explain the decreased root:shoot ratio that occurs when plants are subjected to shorter photosynthetic duration (2, 9).

Studies spanning the entire day/night cycle indicate that partitioning of translocated carbon among sinks changes between day and night (5, 6). At night, accumulation of imported carbon by developing leaves was favored over that by roots in the sugar beet plants studied, fulfilling one of the conditions for the operation of the proposed mechanism. Similarly, developing soybean leaves gained weight throughout the day and night, while roots gained weight during the day and lost weight at night (9). Carbon transport to roots appeared to be lower at night, with a notable portion of the carbon needed for metabolism derived from root stores during the night (9, and references therein).

These data point to a potential mechanism linking increased starch formation rate with decreased root:shoot ratio in soybean plants adapting to shortened days (1, 6, 7, 9). Huber (9) has stated the hypothesis in the following terms. Exported carbon derived from carbohydrate stores in source leaves is used primarily for the growth of the shoot at night. New leaves increase in dry weight both day and night, while the roots import less carbon at night and gain little or even lose weight. Consequently, allocation of newly-fixed carbon to reserve carbohydrate in source leaves is important in determining root:shoot ratio. To evaluate the validity of the working hypothesis, it is necessary to examine the time course of starch storage in association with changes in plant growth. Also, the role of starch in the day/night difference in partition between roots and shoots needs to be verified.

Adjustments to day length also involve morphological changes that are not a direct consequence of partitioning of carbon between root and shoot. Soybean plants raised under 7-h days have thinner leaves, as well as a lower root:shoot ratio than those under 14-h photosynthetic period (2). Leaves of plants raised on a short photosynthetic period supplemented with 7 h of low light, had a specific leaf weight 60% of that of those on 14-h days. In spite of halving of daily irradiance, total leaf area of the SD plants was reduced only 14% below that of the LD plants. Similar compensatory patterns of leaf growth and root:shoot ratio in response to day length were reported for barley (8). Potentially, these changes could result in changed root:shoot ratio.

This research was undertaken to test the proposed role of leaf carbohydrate reserves in regulating partitioning of carbon among sinks during adaptation of plant growth to shortened day length. First, a simple model was developed to illustrate the basic tenets of the hypothesis linking changes in storage of leaf starch to alteration of root:shoot ratio. Next, starch accumulation, net carbon exchange (NCE), leaf area, specific leaf area (SLA), root:shoot ratio, and total dry matter accumulation were measured over a several-week period in sugar beet plants maintained on 14-h days (LD) or adapting after transfer to 8-h days (SD). Finally, the data obtained were used with the model to generate a description of partitioning within plants adjusting to shortened day length. Feasibility of the model was evaluated by comparing values for various model parameters with plant growth measurements.

MATERIALS AND METHODS

<u>Modeling of Partitioning in Sugar Beet Plants</u>. A model was developed to illustrate the hypothesis linking starch storage and partitioning of carbon between roots and shoots (Fig. 1). A general description is given below and a more detailed development is contained in the Appendix.

Variables are defined as follows:

- n   Integer day number. n = 0 is the last day that the plant is on the original night length and refers to the day for initial values of shoot and root carbon.
- $S(n)$ Milligrams carbon in shoots at the end of day n. Values are based on dry weight and the proportion of carbon in $(CH_2O)_M$.
- $R(n)$ Milligrams carbon in roots at the end of day n.
- w1   Proportion of current NCE allocated to starch during the original, longer (14 h) days.
- w2   Proportion of current NCE allocated to starch during the changed, shorter (8 h) days.
- x1   Proportion of current NCE partitioned to shoots under long days.
- x2   Proportion of current NCE partitioned to shoots under new short day length.
- y1   Proportion of current NCE partitioned to roots under long days.
- y2   Proportion of current NCE partitioned to roots under new short days.
- z   Proportion of starch partitioned to shoots under long or short days.
- 1-z   Proportion of starch partitioned to roots under long or short days.

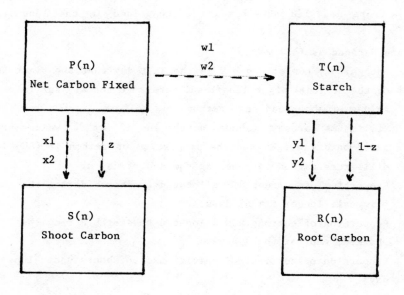

FIG. 1. Model of partitioning in sugar beet plant.

The model was developed with the following aims in mind:

1. To illustrate how changes in the amount of newly-fixed carbon allocated to starch can change the proportions of carbon distributed to roots and shoots and thereby change the root:shoot ratio.

2. To compare the observed kinetics of leaf and plant growth, adjustment of root:shoot ratio and other data from treated and control plants with data generated by the model to evaluate feasibility of the hypothesis.

3. To obtain values for transfer coefficients and other parameters that are instrumental in the adjustment from solution of equations for the model. Experimentally-derived values are used as inputs for this modeling process.

The model is designed to test whether the following two mechanisms are sufficient to explain the change observed in root:shoot ratio:

1. The proportion of newly-fixed carbon allocated to starch increases under short days.

2. The proportion of the total carbon being exported to the shoots increases at night, while that to the roots decreases.

Major assumptions on which the model is based:

1. NCE is computed from a basic rate per area, leaf area per dry weight, leaf dry weight, and the day length. The basic rate is <u>independent</u> of day length.

2. Because a proportion of the carbon partitioned to the shoots and roots is respired, the value for NCE used in the model includes this loss.

3. All newly-fixed carbon that is not allocated to starch is partitioned to shoot and root growth.

4. The coefficient for partition to shoot growth does not distinguish between shoot growth derived from exported carbon and that allocated to growth at the site of NCE.

5. An amount of carbon equivalent to the entire new starch pool is exported by the end of each night.

6. Although the coefficients for partition of newly-fixed carbon to shoots and roots may change with day length, the relative amount partitioned between roots and shoots does not change.

Photoperiod Treatments. Plants of a multigerm variety of <u>Beta vulgaris</u> L., Klein E, were raised under a 14-h photoperiod (25°C, 17°C), under a photon flux density of 450 µmol photons m^{-2} s^{-1} with a 15-min period limited to incandescent illumination at the start and end of each photoperiod. To reduce stress which might be induced by an abrupt day length change (10), the photosynthetic period was shortened by 1 h per day over a 6-d period for the plants transferred to 8-h days. This schedule was adequate to avoid a pronounced disruption of growth (10) beyond what can be expected from reduced daily photosynthetic duration. It appears that the abruptness of the transfer to a new day length may have impact in this type of experimental treatment (10). Four-week-old plants were selected for uniformity, and leaf growth was measured for 8 d. Two plants were kept on 14-h days and four were transferred to 8-h days; leaves were sampled for starch and area measurements continued. The experiment was carried out twice with similar results.

Growth Measurement. Leaf growth was measured by recording the length and width of each leaf daily. Leaf area was obtained from the area of the circumscribing rectangle by means of the leaf:rectangle area ratio determined at the time of harvest. Dry weights of leaf blades, petioles, fibrous roots, and taproot were measured at the end of the treatment period. Leaf disks, 0.17 cm^2 in area, were removed daily, and their dry weight was measured to determine SLA.

Starch Determinations. Four disks with a total area of 0.66 cm^2 were sampled from the 5th leaf of each of the six plants. To obtain starch storage rate, samples were taken at five times throughout each of 10 light periods during the 3 weeks of adjustment to shortened day length. These leaves were ca. 4 plastochrons (10 d) beyond the 3-cm index length when measurements were started. Starch analysis was carried out as described previously (5).

RESULTS

Feasibility of the Proposed Mechanism. The model was used to illustrate the hypothesis linking starch storage to partitioning of carbon between root and shoot (Table I). From the data in Table I, it seems that the mechanism can account for the adjustment in root:shoot ratio

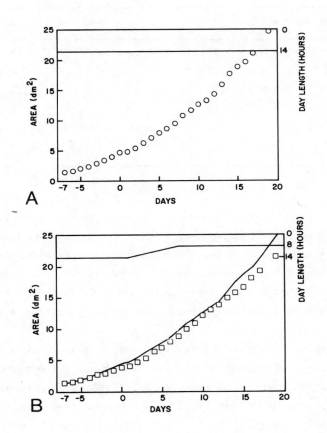

FIG. 2. Growth in leaf area for plants on 14-h and 8-h light days. Data are averages from plants raised under 14-h days. Top curve, four plants grown under 14-h days. Bottom curve, two plants from experiment 2. Line in bottom curve is the curve for long days from top curve.

Table I. *Inputs and Values for Outputs of Carbon Partition Predicted by Partitioning Model*

Values for inputs are data from plants under SD and LD treatments. For allocation of NCE to starch, NCE was the average computed for the treatment period for 14-h and SD plants; starch synthesis in the 5th leaf is the average rate observed for the 14-h plants and the highest average daily rate during transition for SD plants.

Model Inputs			Predicted Partitioning		
Root:shoot ratio	(14 h):	0.39	NCE to shoots	(14 h):	0.61
	(8 h):	0.29		(8 h):	0.49
NCE to starch	(14 h):	0.10	NCE to roots	(14 h):	0.28
	(8 h):	0.28		(8 h):	0.22
			Starch to shoots	(14, 8 h):	0.99
			to roots	(14, 8 h):	0.01

that occurred when the plants were transferred to short days. Values for the transfer coefficients seem reasonable. These data are from the second experiment in which more extensive measurements were made; data from the first experiment likewise confirmed the feasibiity of the mechanism. It should be noted that the new root:shoot ratio, 0.29, is at the end of the allowed range with almost no carbon from starch partitioned to the roots (0.01).

Plant Growth. The six plants used for each experiment were chosen for uniformity from a group of plants. The form of the plants following the 19-d treatment period is summarized in Table II. Leaf growth for LD and for SD plants is shown in Figure 2. Leaf area of the SD plants adjusts without a noticeable lag and slightly exceeds that of the LD plants. In the first experiment, there was a slowing of leaf area growth for 1 to 2 d during the transition period.

Leaf weight was obtained by using the daily samples of leaf SLA for the fifth leaves to convert leaf area to weight. SLA was obtained from one leaf per plant during the course of the experiment. At the end of the

Table II. Comparison of Sugar Beet Plants Adjusting to 8-h Days with Those Remaining on 14-h Days

Measurements are from experiment 2, at end of experiment (19 d).

	Blade			Shoot	Root	Total	Root:shoot
	Dry Wt	Area	SLA	Dry Wt	Dry Wt	Dry Wt	Ratio
	(gm)	(dm^2)	(dm^2 gm^{-1})	(gm)	(gm)	(gm)	
LD:	15.2	24.9	1.62	19.7	17.0	36.7	0.80
SD:	8.9	21.4	2.43	12.7	4.0	16.7	0.29

treatment period, final leaf weights and areas were obtained from each leaf. Leaf weights for the two groups begin to diverge shortly after day length is changed (Fig. 3, day 4). Leaf morphology appears to respond as described in earlier studies (3, 10).

Starch Accumulation. Starch accumulation rate was measured throughout the period of adjustment to shortened days in both treated and in control plants left on 14-h days (Figs. 4 and 5). The failure of the increased rate of starch accumulation to persist was surprising. In both sets of experiments, the compensating increase in starch accumulation expected from reports in the literature lasted for only a few days. Only during the transition (days 6 and 9 in Fig. 4, and 5-7 in Fig. 5) are the rates markedly higher than those of the LD controls. The pattern observed in the first set of experiments was considered in the design of sampling in the 2nd experiment. The 2- to 3-fold increase in rate would have been easily detected given the reliability of the starch accumulation data revealed in the regression analysis.

The results are consistent with our present working hypothesis, that the inital increased rate of starch storage observed in past studies may result, at least in part, from stress caused by the abrupt change in day length. When day length was shortened gradually, growth continued (Figs. 2B and 3) and starch accumulation showed only a transitory increase (Figs. 4 and 5).

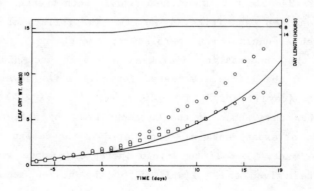

FIG. 3. Average total leaf blade weight for leaves from LD plants and from those (SD) adapting to 8-h days. LD plants (circles) were maintained at 14-h day length, while SD plants (squares) were slowly shifted to 8-h days (see lines at top). Lines are growth in blade weight simulated by the model with data from Table I and the blade:shoot partitioning observed in LD and SD plants.

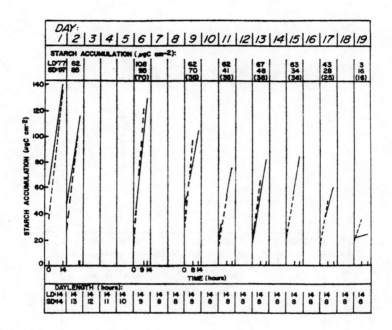

FIG. 4. Daily starch accumulation in leaf 5 of LD plants or SD plants. There were 20 points per regression line for the SD (dashed line) and 10 per line for the LD (solid line) plants; each line had a regression coefficient of at least 0.993. Solid line is omitted where it would obscure the dashed line. Data from each plant (experiment 2) showed the same progression as did the averaged data. Day number indexed from the first day of treatment is given in italics (top of figure). Starch accumulation, shown by the numbers at the top of the graph, is for 14 h (LD), or for the day length shown (at bottom of figure) (SD). The numbers in parentheses give starch accumulation that would have occurred on the shortened day <u>if</u> the rate for the SD plants were the same as for the LD control.

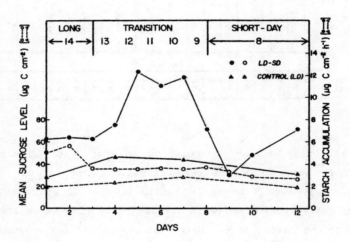

FIG. 5. Sucrose levels and starch accumulation in leaves of SD and LD plants during transition to short days. Data are from experiment 1. Open symbols are sucrose levels while solid symbols are starch accumulation. Plants held at LD are shown by triangles, while plants undergoing a day length transition are shown as circles.

Change in Root-Shoot Partitioning. The data generated by the model demonstrate the feasibility of changing root:shoot ratio by increasing starch storage rate in plants on short days (Fig. 3; Table I). Still, the lack of increased starch storage in plants adapting to short days rules out the mechanism in our experiments. Background for an alternative analysis is presented in Tables III and IV.

Data in Table III show that area available for photosynthesis in the SD plants keeps pace with that of the LD plants (93% of the area of the LD plants at the end of the treatment period). Even though the area may be the same, the leaves are thinner (larger SLA) for SD plants. Multiplying day length by leaf area for each day of the treatment period and summing these gives a cumulative NCE area duration for the period. Dividing the growth in carbon during the treatment period for each set of plants (LD: 33,000; SD: 13,000 mg C) by the cumulative NCE area duration yields the average NCE for LD and SD plants during the period (Table III). In Table IV, the calculated, cumulative NCE during the treatment period

Table III. Cumulative Area Available for NCE During Various Periods in Control Plants and Those Adapting to Shortened Day Lengths

Cumulative photosynthetic area values are the total of the daily leaf areas summed over the period specified. Cumulative NCE area duration given is the sum of the product of leaf area times day length for each day of the 18-d treatment period.

	Cumulative Photosynthetic Area						Cumulative NCE Area Duration		Average NCE	
	Initial 8-d Period		18-d Treatment Period		Total 26-d Period					
	(dm^2)	(%)	(dm^2)	(%)	(dm^2)	%	(dm^2 h)	(%)	(mg C dm^{-2} h^{-1})	(%)
LD:	25.8	100	250.1	100	275.9	100	3502	100	3.8	100
SD:	25.3	98	231.9	93	257.2	93	2048	58	2.5	67

Table IV. Comparison of Carbon Fixed by Sugar Beet Plants Adapting to Short Days With That by Control Plants

The values in parentheses refer to the amount of carbon that would have accumulated in the leaf blades of the SD plants if partitioning remained the same as before the transition to short days. Prediction based on the observed NCE area duration and plant NCE. The first column is the same as the 4th column of Table III. Cumulative NCE is the product of the last two major columns of Table III.

	Cumulative NCE Area Duration Days 1 - 18		Cumulative NCE Days 1 - 18		Leaf Blade Dry Wt					
					Days -7 - 0		Days 1 - 18			
							Observed		Projected	
	(dm^2 h)	(%)	(gm C)	(%)	(gm C)	(%)	(gm C)	(%)	(gm C)	(%)
LD:	3502	100	13.3	100	.50	100	5.3	100	--	--
SD:	2048	58	5.1	38	.46	92	2.9	55	(2.0)	(38)

is shown for both LD and SD plants. The latter is 38% of the former because of lower NCE area duration (shorter days) (8/14 = 57% of the LD plant photosynthetic duration) and lower NCE rate (thinner leaves). The observed difference in blade dry weight is compared with the value predicted from the NCE area duration and NCE rate. The actual gain in blade dry weight is 55% of that in the LD compared with 38% based on the predicated conditions. The extra 0.9 gm likely represents the increase in the partitioning of carbon to shoots that took place during the adjustment to short days.

DISCUSSION

When interpreting starch accumulation data, the trend in starch storage rates as a leaf ages must be considered. Sampling for starch and SLA was begun 4 to 5 plastochrons after the leaf had reached the 3-cm index length, as early as was practical, and was continued until the leaf neared senescence, ca. 10 plastochrons later. Throughout this period, starch accumulation under constant LD conditions remained nearly steady before it diminished to near 0 as the leaf approached senescence (Figs. 4 and 5). The same progression was seen in leaves of the SD plants except for the brief increase during the transition in day length. In both groups, all plants showed the same pattern with no more than a 2-fold difference in starch accumulation rate among plants during the control period. If starch accumulation were going to increase, it would have to occur during our sampling period. From the data, we conclude that only a passing increase in rate occurred, and this always during the transition in day length in the SD plants. The transient shown in Figures 4 and 5 was observed at the same time in all eight SD plants from the two experiments.

These results show that the 2- to 5-fold increase reported by Chatterton and Silvius (3) for a number of species appears not to be a direct and necessary effect of photoperiod on leaves. In fact, the increase was not observed by these authors in a single treated leaf when the other leaves were kept on the original day length (4). These authors concluded that the response was a whole-plant effect. Even the rapid onset of leaf starch accumulation seems to be responding to the growth and morphological development of the plant as a whole.

Observations by Robinson (10) seem to provide a key for understanding the phenomenon of increased starch storage on shortened days. A 15-d lag in plant growth was observed when plants were transferred directly to short days. The magnitude of allocation of newly fixed carbon to starch was always greatest in the earliest stages of exposure to short days. This was interpreted as a result of decreased sink-organ development in SD plants during the adjustment period. Decreased allocation of products of photosynthesis to starch in leaves of the SD plants accompanied the improved growth rate that occurred with time. Comparison of the results of the present study with those of Robinson (10) lead us to conclude that the increase in starch storage immediately after transfer to short days is the result of stress; growth is likely reduced also. The slight, temporary increase in starch storage during the adjustment period, when plants are transferred more gradually, likely mirrors the lesser degree of stress. A 1-d drop in leaf growth was noted during the transition period in the first experiment supporting the existence of some stress. The lag was not visible during the 2nd experiment.

In their studies on the adjustment to shortened days, Chatterton and Silvius (2) were careful to compare their results when plants were transferred to short days with groups of plants kept under either long or short days throughout their life. They note that the rate of starch accumulation shortly after transfer to short days was similar to that in plants grown throughout their life on short days. This rapid adjustment to shortened days was seen as a programmed, whole-plant response meeting the increased need for carbon during the lengthened nights. Huber (9) proposed that the increased availability of carbohydrate reserves in source leaves under short days contributes to the increase in root:shoot ratio. The idea that increased allocation of newly-fixed carbon to starch is a means for changing root:shoot ratio is an attractive one (5, 6). It now appears that the long-term increase in starch accumulation in SD plants may be a result of the integrated response of the plant to carbon use on short days rather than a direct effect on the source leaf. The day length signal does not change allocation of recently-fixed carbon directly. The impact of day length on plant growth and morphology results in changes in allocation and partitioning of carbon that are part of the integrated response to changes in growth. Cause/effect relations differ from those that would obtain if starch accumulation were a direct programmed response in the source leaves.

Future Studies. This report deals with the initial findings of this study. Further testing of the hypothesis that increased starch storage is the result of an effect on sink region growth is currently underway.

Acknowledgments--This work was supported by NSF Grant DMB-8303957 and grants from the University of Dayton Research Council and Shell Development Company. We gratefully acknowledge the skillful assistance of Mary Kolesnicky, Mary Komorowski, Kelli McCafferty, Mark Palmert, Walter Reiling, and Angela Stevens.

ERRATA:

Line 12 on page 304 should read

"...decrease in root:shoot ratio" rather than ...increase...

LITERATURE CITED

1. CHATTERTON NJ, JE SILVIUS 1979 Photosynthate partitioning into starch in soybean leaves I. Effects of photoperiod versus photosynthetic duration. Plant Physiol 64:749-753
2. CHATTERTON NJ, JE SILVIUS 1980 Acclimation of photosynthate partitioning and photosynthetic rates to changes in length of the daily photoperiod. Ann Bot 46:739-745
3. CHATTERTON NJ, JE SILVIUS 1980 Photosynthate partitioning into leaf starch as affected by daily photosynthetic period duration in six species. Physiol Plant 49:141-144
4. CHATTERTON NJ, JE SILVIUS 1981 Photosynthate partitioning into starch in soybean leaves II. Irradiance levels and daily photosynthetic period duration effects. Plant Physiol 67:257-260
5. FONDY BR, DR GEIGER 1982 Diurnal pattern of translocation and carbohydrate metabolism in source leaves of Beta vulgaris L. Plant Physiol 70:671-676
6. FONDY BR, DR GEIGER 1983 Control of export and partitioning among sinks by allocation of products of photosynthesis in source leaves. In D Randall, D Blevins, R Larson, B Rapp, eds, Current Topics in Plant Biochemistry and Physiology. University of Missouri, Columbia
7. GEIGER DR, BR FONDY 1985 Programmed responses of export and partitioning to internal and external factors. In B Jeffcoat, ed, Regulation of Sources and Sinks in Crop Plants, Monograph 12. British Plant Growth Regulatory Group
8. GORDON AJ, GJA RYLE, DF MITCHELL, CE POWELL 1982 The dynamics of carbon supply from leaves of barley plants grown in long or short days. J Exp Bot 33:241-250
9. HUBER SC 1983 Relation between photosynthetic starch formation and dry-weight partitioning between the shoot and root. Can J Bot 61:2709-2716
10. ROBINSON JM 1984 Photosynthetic carbon metabolism in leaves and isolated chloroplasts from spinach plants grown under short and intermediate photosynthetic periods. Plant Physiol 75:397-409

Appendix: Supplementary Documentation for Model

<u>Additional Variables</u>: The following additional variables are used in computations (1 refers to original, LD condition and 2 refers to new, SD condition):

$Q(n)$ root:shoot ratio at the end of day n. $Q(n) = R(n)/S(n)$.
$Q1$ initial root:shoot ratio at the end of day n. $Q(n) = R(n)/S(n)$.
$Q2$ final root:shoot ratio under the new day length.
$P(n)$ milligrams carbon added to plant during day n by net photosynthesis. The latter is extended to mean true photosynthesis, less loss by both photorespiration in the leaves and dark respiration by roots and shoots throughout the 24-h period of day n.
$T(n)$ milligrams carbon in starch storage at the end of the daylight period of day n. We assume that this is all transported during the night period.
c milligrams carbon added per hour to plant by net photosynthesis in 1 cm^2 leaf area, assumed to be independent of day length.
a leaf area density; leaf area for a unit weight shoot carbon.
Ln number of daylight hours in day n.

<u>Hypotheses to be Tested</u>.
1. $w1 < w2$ days   Allocation of NCE to starch increases under short days.
2. $(1-z)/z < y/x$ The ratio of root:shoot partitioning is smaller at night than during the day.

<u>Equations of the Model</u>. From our assumptions about NCE.
(1) $P(n) = c * Ln * a * S(n-1)$
(2) $= J * S(n-1)$ where $J = a * c * Ln$, cumulative NCE.

Thus, it follows that

$T(n) = w * P(n)$
$S(n) = S(n-1) + x * P(n) + z * T(n)$
$R(n) = R(n-1) + y * P(n) + (1-z) * T(n)$

Substituting for $P(n)$ and $T(n)$ and solving the resulting recurrence relations yields:

(3) $S(n) = (1 + x * J + z * w * J)^n * S(0)$
          $= H^n * S(0)$ where
   $H = 1 + x * J + z * w * J$

(4) $R(n) = R(0) + (y + (1-z) * w) * J * S(0) * ((H^n-1)/(H-1))$

The root:shoot ratio at the end of day n is

$Q(n) = R(n)/S(n)$ and substituting gives:

(5) $Q(n) = R(0)/(S(0) * H^n) + (y + (1-z) * w) * J * ((1-H^{-n})/(H-1))$

We can find the equilibrium root:shoot ratio, Q, by evaluating the limit of $Q(n)$ as n --> infinity, and where $H^n \gg 1$.

(6) $Q = (Y + (1-z) * w)/(x + z * w)$

The values of $S(n)$, $R(n)$, $Q(n)$, and Q found in equations (3) to (6) are valid for a plant which has come to equilibrium while being grown under a given, fixed day length.

Let us assume that a plant is initially grown under the condition of 14-h day length. In this case, its carbon partitioning coefficients are w1, x1, y1, z, and 1-z, and the equilibrium root shoot ratio is Q1. We then assume the day length is changed to 8 h and a new equilibrium is approached. Here, the partitioning coefficients are w2, x2, y2, z, and (1-z) with the limiting root:shoot ratio being Q2. Now, experimentally, we can measure Q1, Q2, w1, and w2. We wish to evaluate the remaining partitioning coefficients using these values and the assumptions of the model. From equation (6)

$Q1 = (y1 + (1 - z) * w1)/(x1 + z * w1)$

Solving for z

(7) $z = (B1 - w1)/w1$ where $B1 = 1/(1 + Q1)$

Similarly, evaluating Q2 and solving for z yields

(8) $z = (B2 - x2)/w2$ where $B2 = 1/(1 + Q2)$

From the assumptions of the model, we have three other equations:

(9) $x1 + y1 + w1 = 1$

(10) $x2 + y2 + w2 = 1$

(11) $x1/y1 = x2/y2$

Solving simultaneously and setting $D = (w2 * B1 - w1 * B2)/(w2 - w1)$, we have:

(12) $x1 = (1-w1) * D$

(13) $y1 = (1-w1) * (1-D)$

(14) $x2 = (1-w2) * D$

(15) $y2 = (1-w2) * (1-D)$

(16) $z = (B1 - (1-w1) * D)/w1$

Though these numbers are mathematically well-defined, provided that $w2$ is not equal to $w1$ and $w1$ is not equal to 0, they are biologically well-defined only if all the partitioning coefficients are values between 0 and 1. Hence, we conclude:

(17) $\quad 0 < D < 1$

(18) $\quad 0 < (B1 - (1-w1) * D)/(w2 - w1) < 1$

which implies that once any three of these values $Q1$, $Q2$, $w1$, and $w2$ have been fixed, the last, while not completely determined, is nonetheless limited in value.

Given values for $w1$, $w2$, and $Q1$, and using the convention $w1 < w2$, the constraints for $Q2$ are given by:

(1) $\quad Q2 < ((1-w1) * (1 + Q1)/(1 - w2)) - 1$

(2) $\quad Q2 > ((1-w2) * Q1)/((1 - w1) + (w2 - w1) * Q1)$

(3) $\quad Q2 > (w1 * (1 + Q1)/w2) - 1$

(4a) If $Q1 < w1/(w2 - w1)$ then $Q2 < (w2 + Q1)/(w1 - (w2 - w1) * Q)$

(5) If $Q1 \geq w1/(w2 - w1)$ then $Q2 \geq 0$ is the only restriction.

Other Carbohydrates as Translocated Carbon Sources: Acyclic Polyols and Photosynthetic Carbon Metabolism

WAYNE H. LOESCHER, JOHN K. FELLMAN,
TED C. FOX, JEANINE M. DAVIS,
ROBERT J. REDGWELL, and ROBERT A. KENNEDY

It was at a symposium like this that someone once made the plea that any proper study of carbohydrate metabolism should begin with sucrose. If so, then the topic of this presentation is decidedly improper, for an extensive discussion of the acyclic polyols (sugar alcohols, polyalcohols or polyhydric alcohols) need not mention sucrose at all. Although acyclic polyols form a broad group of compounds that closely resemble the sugars (chemically, physically, and biologically) from which they are derived, they are also, in many ways, unique.

Polyols are widely distributed throughout the plant kingdom. One reviewer (5) suggests that for those interested in carbohydrate metabolism in plants, it would be wise to assume that a polyol could be present until shown otherwise. They are common constituents of fungi, algae, and lichens, and frequently constitute the major carbohydrate present (21). As major constituents of the Phaeophyta and Chrysophyta, as well as in most fungi, polyols contribute significantly to the world's carbon economy. Bieleski (5) estimated that about 30% of the annual global primary production goes through polyols rather than sugars, excluding oxidation back to CO_2 through rots and decays where the fungi are undoubtedly major agents. In angiosperms, 13 polyols have been isolated. Of these, mannitol is perhaps the most widely distributed, having been found in over 70 higher plant families. Though less widespread, other polyols are common in various angiosperm taxa. Dulcitol (galactitol), for example, is characteristically found only in the Celastraceae. Sorbitol (glucitol), although present in a number of other higher plants, is found most often in the

Rosaceae, where it is a common constituent of many economically important crop plants (31).

Probable physiological functions of polyols include storage of carbohydrates and reducing power (21, 47), regulation of co-enzymes (43), involvement in osmoregulation (13) and use as compatible solutes (6, 55). Evidence for such physiological roles comes primarily from studies on fungi and animals. Very little work has been done to determine physiological roles of polyols in any green plant. It is known, however, that along with sugars and nonreducing oligosaccharides, polyols share the role of being the form in which carbon is translocated in the phloem (56, 57). Polyols play a similar role in some algae (29). Further, in many algae, lichens, and some families of higher plants, polyols are the major products of photosynthesis (1, 29, 48), replacing neutral sugars such as sucrose. Sorbitol, for example, is the major product of photosynthesis in many species in the Rosaceae, including all the members of the economically important genera <u>Malus</u> (apples), <u>Pyrus</u> (pears), and <u>Prunus</u> (stone fruits such as peach, cherry, plum, and apricot). In these species, sorbitol accounts for 60 to 90% of the carbon exported from the leaf (50). Mannitol plays a similar role in many important members of the families Scrophulariaceae (snapdragon), Apiaceae (celery), Rubiaceae (coffee), and Oleaceae (olive) (5, 49). Polyols also serve as storage compounds in organs of some higher plants. Sorbitol is found in many common fruits. In other tissues, seasonal variations have been reported in some species, and several investigators have correlated sorbitol accumulation with frost hardiness (9, 30, 35, 41). Similar data have been reported for members of the Oleaceae, where mannitol is the polyol involved in translocation and storage. Another role for polyols is as osmotica. This has been demonstrated in several mangrove species (32), and there is clear evidence for osmoregulatory interconversions of starch and sorbitol in <u>Plantago</u> (8).

Only in the past five years has much progress been made on metabolism of the acyclic polyols in higher plants. We now know that synthesis of sorbitol proceeds via NADPH-dependent aldose 6-phosphate reductase which catalyzes the reaction (15, 28):

$$\text{Glc-6P} + \text{NADPH} \rightleftharpoons \text{sorbitol-6-P} + \text{NADP} \qquad [1]$$

This enzyme, plus a specific phosphatase (10), constitutes sorbitol synthesis in source tissues. Sink tissues such as young meristematic leaves (22) contain a NAD-dependent sorbitol dehydrogenase which catalyzes the reaction (27):

$$\text{sorbitol} + \text{NAD} \rightleftharpoons \text{fructose} + \text{NADH} \qquad [2]$$

This scheme does not, however, account for reports of sorbitol oxidase (53), NADPH-dependent sorbitol dehydrogenase (14, 54) or sorbitol-1-phosphate (37). Additional work has shown that celery and other mannitol-synthesizing higher plants contain a NADPH-dependent mannose-6-phosphate reductase (MPRase) which catalyzes the reaction (23):

$$\text{mannose-6-P} + \text{NADPH} \rightleftharpoons \text{mannitol-1-P} + \text{NADP}^+ \qquad [3]$$

This enzyme is cytosolic along with mannose-6-phosphate isomerase, mannitol-1-phosphate phosphatase and NADP-linked, nonreversible glyceraldehyde-3-phosphate dehydrogenase (40).

The information now available provides insight on acyclic polyol metabolism in higher plant tissues, but there is little to explain why these compounds have arisen as important metabolites in so many different taxa. Both reducing and nonreducing sugars fulfill all the roles described above in many other taxa. Another explanation may be the stoichiometry of the reactions involved in the synthesis of acyclic polyols. These processes require utilization of up to one-third of the triose-phosphate (triose-P) exported from the chloroplast for generation of reductant. Recycling of reduced pyridine nucleotides may provide several advantages to the plant. The reductive pentose phosphate pathway is only one of several sets of reactions in chloroplasts which result in, or require, recycling pyridine nucleotides. Reduced pyridine nucleotides are produced in chloroplasts in the dark during starch breakdown via glycolysis or the oxidative pentose phosphate pathway (17, 44, 45). Synthesis and export of polyol during starch breakdown would be one mechanism of recycling reductant. In light, recycling pyridine nucleotides would be advantageous to avoid oxidative photodestruction or photoinhibition, as has been postulated by several investigators (33). These stresses occur when CO_2 is limiting or when unacclimated leaves are exposed to high light intensity. Other reactions which, in essence, recycle pyridine nucleotides are the Mehler-type reactions (26, 34, 51), as well as photorespiration.

A major explanation for the occurrence of photorespiration is that the process serves to waste excess NADPH that cannot be used for CO_2 fixation or other growth processes (46). Due to photorespiration, electron transport and the associated photochemistry may proceed continuously in light and yet never completely reduce the available pool of pyridine nucleotides. Consequently, polyol synthesis may represent another means of accomplishing the same purpose, or alternatively, utilizing otherwise "wasted" reducing power. Observations we and others have made may be pertinent here. Photosynthetic rates in young, fully-expanded apple leaves are generally very high (3), 40 to 43 mg CO_2 dm^{-1} h^{-1} (16), and substantially higher than what is thought typical of trees (19) or many C_3 plants. Data also indicate that mesophyll resistance is quite low in apple leaves (16). Whether these observations are related to a polyol being the principal photosynthetic product in apples remains to be determined. It may also be pertinent to note that in Larcher's 1969 review and survey of photosynthesis in trees (19), among the highest values reported are for Pyrus, a sorbitol synthesizer, and Fraxinus, a mannitol synthesizer. Unfortunately, there are very few photosynthetic studies of polyol-synthesizing species. With the exception of apple, none that we are aware of has been studied in any detail.

As a consequence of these and other observations, we have set out to measure photosynthesis and carbon metabolism in celery, Apium graveolens, a mannitol synthesizer. Celery is a convenient model for such work for several reasons, not the least of which is that it can readily be grown in conventional growth chambers. It also synthesizes, in addition to mannitol, both starch and sucrose, thus offering a convenient system for study of partitioning between those photosynthetic products. In addition, its growth habit provides a continuing supply of leaves at different developmental stages, a desirable characteristic for studies on sink-source transitions. Most importantly, when its exacting environmental requirements are met, the plant is capable of tremendous biomass production, up to 135 metric tons/ha (11). This alone indicates that its photosynthetic rates deserve close study. Our objective has been to determine, through measurements of gas exchange and carbon metabolism, what may be photosynthetically unique in a plant that we already know to be highly productive of both biomass and polyol.

MATERIALS AND METHODS

Plant Material. Celery plants (<u>Apium</u> <u>graveolens</u> L. 'Giant Pascal') were grown in a local greenhouse with supplemental lighting under optimal conditions that were previously determined.

Photosynthesis Measurements. The three terminal leaflets of a leaf were sealed in a Plexiglass chamber, and photosynthetic rates were determined using a Beckman Model 865 infrared gas analyzer in an open system (16). Oxygen and relative humidity readings were made with a Beckman 0265 oxygen analyzer and a WeatherMeasure HM111RG relative humidity sensor, respectively. Measurements were performed at a PPFD of 1200 to 1800 μmol photons m^{-2} s^{-1} provided by halide lamps. Leaves not being measured were shaded to prevent desiccation. Leaf area measurements were taken with a LiCor Model 3100 leaf area meter. Carbon dioxide compensation points were determined in a closed system. CO_2 was removed from the system with Ascarite and individual leaves were sealed in a 125-ml glass tube containing 10 ml deionized water and held at 25°C in a water bath at irradiance levels of 1200 μmol photons m^{-2} s^{-1}.

Incorporation of $^{14}CO_2$ and Fractionation of Metabolites. Fully expanded, attached terminal leaves of similar age were enclosed in a plastic bag after which 50 μCi of $NaH^{14}CO_3$ and 100 μl of 30% lactic acid were introduced via a gelatin capsule placed in the bag and ruptured at time zero. Plants were allowed to photosynthesize for various times between two mercury vapor lamps at 27 to 28°C, after which plant material was freeze-clamped between aluminum blocks previously equilibrated in liquid nitrogen. All tissues were extracted and fractionated into neutral (sugar), amino acid, organic acid, and phosphate ester moieties (36).

Preparation of Enzyme Extracts. Upon transfer to the laboratory, 11 to 12 leaf plants (usually 10-12 weeks old) were prepared for most enzyme assays in the following fashion: leaf punches (1.0 cm^2, 2x) were taken from leaves of different ages for subsequent chlorophyll analysis; 1.0 g (fresh weight) of leaf tissue was ground in a pre-chilled mortar and pestle containing 0.1% w/v PVP and acid-washed quartz sand in buffer composed of 50 mM Hepes-NaOH, 1 mM EDTA-Na$_2$, 5 mM MgCl$_2$, 1 mM DTT at pH 7.0. After grinding, the brei was transferred to microcentrifuge tubes and centrifuged at 12,800g for 1 min to remove sand and PVP. The supernatant was transferred to another microcentrifuge tube and re-centrifuged at 12,800g for 4 min. The supernatant from the second centrifugation was used for all enzyme assays except MPRase. For assay of MPRase, 0.2 g fresh weight plant material was ground with sand and PVP in 100 mM Tris buffer containing 10 mM DTT pH 7.5, centrifuged at 12,800g for 1.5 min, and allowed to stand on ice for 10 min. Aliquots of the supernatant were then used for enzyme analysis.

Enzyme Assays. Ribulose-1,5-bisphosphate carboxylase (Rubisco, E.C. 4.1.1.39) was assayed using the method of McFadden and Denend (25). Alkaline fructose-1,6-bisphosphate,1-phosphatase (FbPase, E.C. 3.1.3.11) was assayed according to Latzko and Gibbs (20). MPRase activity was determined using the procedure described by Rumpho et al. (40). Sucrose phosphate synthase (SPSase, E.C. 2.4.1.14) and sucrose synthase (SSase, E.C. 2.4.1.13) were assayed by a modification of the procedures of Rufty et al. (39) using glass-capillary, gas chromatography with flame ionization detection of the sucrose synthesized in the presence of UDPG and fructose-6-phosphate in the case of SPSase or fructose in the case of SSase. Phosphoenolpyruvate carboxylase activity (PEP carboxylase, E.C. 4.1.1.31) was measured using the method of Williams and Kennedy (52), following incorporation of H^{14}CO$_3$ into acid-stable products in the presence of PEP and glucose-6-phosphate.

Chlorophyll and Protein Analyses. Total chlorophyll was determined using the procedure of Arnon (2). Protein was measured according to Bradford (7).

RESULTS

Although celery's productivity on a biomass per unit land area basis is well documented (11), we are not aware of any reports of photosynthetic gas exchange. Our data show that celery plants possess a very large capacity for photosynthesis. They exhibit higher rates than are commonly found in most C$_3$ plants, either on a chlorophyll or leaf area basis.

Developmentally, the highest rates were seen in young, vigorously-growing plants; in plants with nine leaves, the highest rates (Fig. 1A) were found in leaf 8 (leaf position corresponds with leaf age, with leaf number one being the youngest visible in the spiral). Mature plants with 19 leaves exhibited maximum photosynthetic rates in leaves 12 to 16 (Fig. 2B). Only in the oldest leaves did photosynthetic capacity decline. Unlike other species where maximum photosynthetic capacity has been reported to occur when leaves attain 70 to 100% of full leaf expansion (24), celery appear to maintain maximal photosynthetic rates longer. The high rates, however, were typical only of vigorous plants; subjecting plants to transient environmental stress (water, nutrient, or temperature stress) did not result in enhanced photosynthetic rates in either existing or subsequently formed leaves (data not shown). The response of photosynthesis to temperature was typical of C_3 plants: the temperature optimum occurred over a broad range (22-28°C), peaking at 26°C. Light saturation occurred at relatively low values, 600 μmol photons m^{-2} s^{-1} for all leaf ages studied (data not shown). Although these CO_2 exchange data may be, in part, the consequence of optimum growing conditions, they are also consistent with the field data reporting exceptionally high rates of biomass production on a unit land area basis (11).

In addition to photosynthesis, photorespiration also changed with leaf development, as seen by the photosynthetic response at 2 and 21% O_2. In the four stages of leaf development shown in Table I, the degree of photosynthetic enhancement (2% O_2 compared to 31% O_2) increased, suggesting an increase in photorespiration with leaf age. Similarly, the CO_2 compensation points increased as leaf age increased; values ranged from 2.4 μl/L for leaf 3 to 30.0 μl/L for leaf 13. These CO_2 compensation points are remarkably low for a C_3 plant. For comparative purposes, data on CO_2 compensation points of soybean and corn were obtained in our laboratory using the same methods as were used for celery, and the values for both species were typical of those C_3 and C_4 plants. In celery, carboxylation efficiency (CE) increased and mesophyll resistance (R_m) decreased with leaf age (Table I). These data are similar to the pattern observed in apple (16), a sorbitol synthesizer, but the values for CE are lower and R_m is higher in celery.

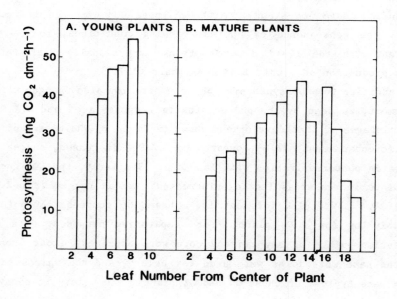

FIG. 1. Effects of leaf and plant age on photosynthetic rates in young (A) and mature (B) celery plants.

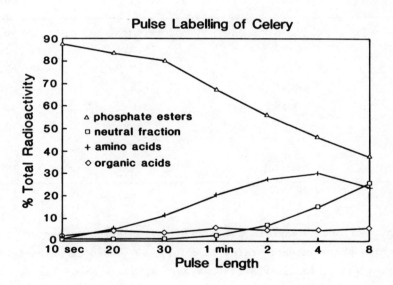

FIG. 2. Incorporation patterns for pulse-labeled $^{14}CO_2$ in young, fully-expanded celery leaves. See "Materials and Methods."

Table 1

Photosynthetic enhancement, carboxylation efficiency, mesophyll resistance to CO_2 diffusion and CO_2 compensation points at four stages of leaf development in celery plants.

Leaf Number	Photosynthetic Rate		Enhancement (%)
	21% O_2	2% O_2	
	(mg CO_2 dm^{-2} h^{-1} at 300 ppm CO_2)		
3	18.4	21.7	18.0
4	37.1	43.9	18.4
6	27.9	34.6	23.9
8	45.7	59.6	23.3

Leaf Number	Carboxylation Efficiency (CE)	Mesophyll Resistance (R_m)	CO_2 Compensation Point
		(s/cm)	($\mu l/l$)
3	0.07	15.2	2.4
4	0.14	7.0	2.7
6	0.13	7.9	3.1
8	0.18	5.6	9.2
13	-	-	30.0
Corn	-	-	1.5
Soybean	-	-	42.1

In an effort to explain the low CO_2 compensation points and the relatively high photosynthetic rates in celery, we conducted pulse-labeling of $^{14}CO_2$ into young, fully-expanded leaves. Although these leaves typically had low CO_2 compensation points, Figure 2 shows that incorporation patterns were typical of a C_3 plant with the bulk of the radioactivity initially in the phosphate ester fraction, followed later by a rise in the organic and amino acid fractions and ultimately by accumulation primarily in the neutral sugars. There was little label in the insoluble fractions following either pulse or pulse-chase incorporations, with the pulse-chase data indicating that most (> 90%) of the label ultimately appeared as mannitol and sucrose (data not shown).

In another effort to explain the low CO_2 compensation points, we measured PEP carboxylase activities in crude extracts. Figure 3 shows the results for leaves of various ages expressed as per cent of maximum activity within the canopy on the basis of fresh weight, protein, and chlorophyll. Despite some change in activity as leaves initially expanded in size, we did not find any appreciable enzyme activity. The maximum celery values (0.021 U g^{-1} fresh weight, 0.0188 U mg^{-1} Chl, 0.007 U mg^{-1} protein, U = μmol min^{-1}) were consistently characteristic of C_3 species (42).

We also measured Rubisco activities in crude extracts from leaves of different ages. Figure 4 shows the results, again expressed as per cent of maximum activity. Regardless of how expressed, Rubisco activities generally increased with the initial increase in leaf size, and then were relatively stable until senescence, except when calculated on a chlorophyll basis. On that basis, there was the initial increase followed by a decrease and then a second increase before the final decline. The second increase was related to the onset of the decline in chlorophyll content as leaves began senescing. For celery, the average maximum activities for Rubisco, 3.46 U g^{-1} fresh weight, 1.88 U mg^{-1} Chl and 0.83 U mg^{-1} protein, appear to be consistent with enzyme activities obtained by workers using spinach (38) and soybean (39). The celery activities may in fact be somewhat higher, but direct comparisons are difficult.

Because of its involvement and apparent regulation in the reductive pentose phosphate pathway, as well as in synthesis of starch, sucrose, and mannitol, we also measured FbPase in crude extracts from leaves at different developmental stages. FbPase activities were invariably quite low in very young, sink leaves, but activities increased dramatically, especially on a fresh weight basis, as leaves matured (Figure 5). Average maximum values were 1.31 U g^{-1} fresh weight, 0.70 U mg^{-1} Chl and 0.43 U mg^{-1}

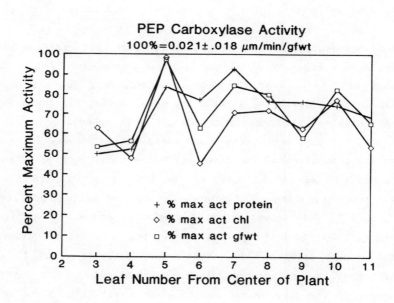

FIG. 3. Effects of leaf age on phosphoenolpyruvate carboxylase activities in crude extracts from celery leaves. All activities are expressed as per cent of maximum activity (of all leaves tested) on a basis of fresh weight (0.021 U g^{-1} fresh weight), chlorophyll (0.019 U mg^{-1} Chl), or protein (0.007 U mg^{-1} protein). Each point is the mean of measurements of three leaves from separate plants.

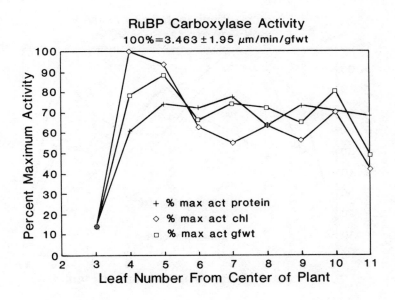

FIG. 4. Effects of leaf age on ribulose bisphosphate carboxylase activities in crude extracts from celery leaves. All activities are expressed as per cent of maximum activity (of all leaves tested) on a basis of fresh weight (3.46 U g^{-1} fresh weight), chlorophyll (1.88 U mg^{-1} Chl), or protein (0.83 U mg^{-1} protein). Each point is the mean of measurements of three leaves from separate plants.

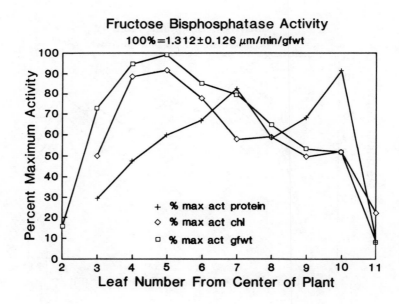

FIG. 5. Effects of leaf age on fructose-1,6-bisphosphate-1,phosphatase in crude extracts from celery leaves. All activities are expressed as per cent of maximum activity (of all leaves tested) on a basis of fresh weight (1.31 U g^{-1} fresh weight), chlorophyll (0.70 U mg^{-1} Chl), and protein (0.43 U mg^{-1} protein) basis. Each point is the mean of measurements of three leaves from separate plants.

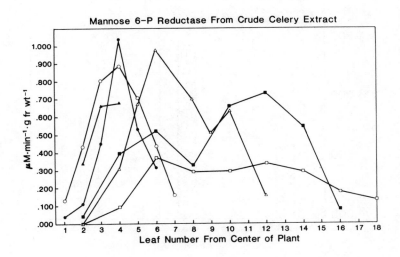

FIG. 6. Activities of mannose-6-phosphate reductase in leaves from plants of different ages. Lines connect leaves from the same plant. Activities expressed on fresh weight (U g^{-1} fresh weight) basis.

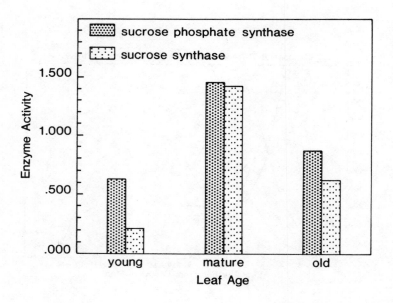

FIG. 7. Activities of sucrose phosphate synthase and sucrose synthase as a function of leaf age. Enzyme activity is expressed as μmol sucrose-6-P or sucrose formed per minute per gram fresh weight.

protein. These values are all consistent with those reported for crude extracts from other species (38, 39).

Generally, comparison of these changes in enzyme activity (Figs. 3-5) with increased photosynthetic rates in maturing leaves (Fig. 1) shows a good relationship between capacity for photosynthesis and both Rubisco and FbPase, but no correlation with PEP carboxylase. On the other hand, although the relatively high Rubisco values may contribute to the high photosynthetic rates observed in celery leaves at some stages of their development, there are no data to explain the exceptionally low CO_2 compensation points. These compensation points are exceptional because both enzyme assays and labeling data (i.e. the low PEP carboxylase activity and lack of appreciable incorporation into organic acids) indicate that celery is a C_3 plant.

We have done some preliminary anatomical studies. Young leaves that possessed the lowest CO_2 compensation points, although they had a compact mesophyll, did not appear to have a well-developed bundle sheath. These observations confirm earlier anatomical investigations done at the light microscopy level (12). At the ultrastructural level, we have not yet investigated the possibility that there may be some peculiar orientation of the chloroplasts and other organelles which might account for internal recycling of photorespiratory CO_2.

To determine if there might be a relationship between onset of high rates of photosynthesis and polyol metabolism, we used MPRase activity as an indicator of capacity for polyol synthesis. MPRase is the key enzyme in mannitol synthesis in celery leaves (reaction [3]). Generally, MPRase activity changed dramatically with leaf age, increasing with photosynthesis, especially as young leaves underwent the sink-to-source transition (Fig. 6). Highest activities were found in the same leaves that were photosynthesizing most rapidly, regardless of whether we assayed leaves from young or more mature plants. For comparison, we assayed SPSase and SSase in bulked samples of young, mature, and old leaves (leaf positions 1-5, 6-10, and >10, respectively) from similar plants (Fig. 7). Although not directly comparable (both the extraction procedures and the plants were different), activities of the sucrose enzymes paralleled both photosynthetic rates and MPRase activities.

DISCUSSION

Celery certainly presents us with an anomaly when we consider the information currently available on its photosynthesis and carbon metabolism. The plant is capable of high photosynthetic rates in some leaves, and these rates are exceptionally high when considered in the context of the rates usually observed in C_3 plants. Such high rates must contribute, at least in part, to the plant's productivity as a vegetable crop. Crop productivity may also be due to an ability, as leaves mature, to maintain maximum photosynthetic rates longer, a trait especially important in older plants where a large fraction of the canopy consists of mature leaves. Neither of these photosynthetic traits, however, can explain the exceptional CO_2 compensation points. Conventional explanations for low compensation points usually include a role for PEP carboxylase, and yet we could find only low levels of this enzyme in celery leaf tissues, regardless of age. The labeling data, which indicate low levels of incorporation into the organic acids also typical of C_3 plants, provide no evidence that there might be an alternative to the reductive pentose phosphate pathway operating in these tissues. Although Rubisco levels in our crude extracts were relatively high, they were not exceptionally high and do not appear to contribute to the low compensation points.

Thus, if we attempt to explain the photosynthetic characteristics of celery on the basis of our present knowledge of carbon metabolism, it appears that the plant truly is an anomaly. Generally, it possesses most of the characteristics of a conventional C_3 plant. The only unusual aspect of its carbon metabolism appears to be mannitol synthesis. Mannitol synthesis, however, has a great deal in common with sucrose synthesis. Synthesis is cytoplasmic for both of these translocatable substances and both appear to be synthesized from the same pool of hexose phosphates, which in turn are generated from the triose-P transported across the chloroplast envelope. Therefore, it does not seem likely that presence or absence of the glycolate pathway provides any direct advantages to the synthesis of either sucrose or mannitol.

On the other hand, mannitol synthesis does involve recycling of pyridine nucleotides. Although the importance of this aspect of acyclic polyol metabolism has been demonstrated in fungi (43), there is as yet no evidence for any such role in higher plant tissues. There is, however, one report (18) of altered respiratory metabolism in <u>Plantago coronopsis</u>. This species osmoregulates by interconversions of starch and sorbitol, and when roots are exposed to high salinity, increased sorbital synthesis is accompanied by decreased alternate respiration. There is also the idea that photorespiration dissipates excess energy produced photochemically and thereby prevents photoinhibition of CO_2 fixation resulting from photooxidative destruction of the photochemical apparatus (33). Illuminating photosynthetic cells under nitrogen in the absence of CO_2 does cause irreversible damage to their capacity for CO_2 fixation. Related to this is the idea that photorespiration could protect the plant from destructive injury when the plant is experiencing water or high temperature stress, periods when internal CO_2 levels would be most limiting. Mannitol synthesis would be an alternative mechanism for dissipation of reductant and it is, therefore, conceivable that the need for photorespiration might be reduced. No matter how desirable this hypothesis might be, the data that we have presented here unfortunately provide, at best, only indirect support for this hypothesis. In addition, we currently have no evidence to explain the low CO_2 compensation points, for example, CO_2 concentrating mechanisms, bundle sheaths, or other structural modifications.

We do propose a set of experiments which are designed to test one possibility. If mannitol synthesis represents a means for utilization of excess reductant, then labeling studies of plants placed under environmental conditions designed to create high levels of reductant should be productive. Pulse and pulse-chase studies of celery plants exposed to high light intensities (above the saturation point) or to subambient levels of CO_2 should allow comparing the per cent of radioactive label in sucrose and mannitol with that of appropriate controls. If the fraction of label in mannitol increases under photoinhibitory conditions, this would indicate that synthesis of polyol was sensitive to availability of reductant, as well as hexose phosphate. Celery is advantageous to use here because it synthesizes both sucrose and starch in addition to mannitol. This would, therefore, provide insight into what controls carbon partitioning among these products of photosynthesis.

Carbon metabolism and photosynthesis have been studied so rarely in polyol plants that we have no idea how widespread what we see in celery may be. Thus, we think it is, now more than ever, entirely proper to make the plea that these subjects be studied in detail in a variety of polyol-synthesizing species.

LITERATURE CITED

1. ANDERSON JD, P ANDREWS, L HOUGH 1961 The biosynthesis and metabolism of polyols. Sorbitol (D-glucitol) of plum leaves. Biochem J 81:149-154
2. ARNON DI 1949 Copper enzymes in isolated chloroplasts. Polyphenol oxidase in Beta vulgaris. Plant Physiol 24:1-15
3. AVERY DJ 1977 Maximum photosynthetic rate--a case study in apple. New Phytol 78:55-63
4. BIELESKI RL 1969 Accumulation and translocation of sorbitol in apple phloem. Aust J Biol Sci 22:611-620
5. BIELESKI RL 1982 Sugar alcohols. In FA Loewus, W Tanner, eds, Encyclopedia of Plant Physiol, Vol 13A, New Series. Springer-Verlag, New York, pp 158-192
6. BOROWITZKA LJ 1981 Solute accumulation and regulation of cell water activity. In LG Paleg, D Aspinal, eds, The Physiology and Biochemistry of Drought Resistance in Plants. Academic Press, New York, pp 97-130
7. BRADFORD MM 1976 A rapid and sensitive method for the quantitation of microgram quantities of protein using the principle of protein-dye binding. Anal Biochem 72:248-254
8. BRIENS M, F LARHER 1983 Sorbitol accumulation in Plantaginaceae; further evidence for a function in stress tolerance. Z Pflanzenphyiol 110:447-458
9. DOWLER WM, FD KING 1966 Seasonal changes in starch and soluble sugars content of dormant peach tissues. Proc Am Soc Hort Sci 89:80-84
10. GRANT CR, T AP REES 1981 Sorbitol metabolism by apple seedlings. Phytochemistry 20:1505-1511
11. HARTMAN HT, WJ FLOCKER, AM KOFRANEK 1981 Plant Science. Prentice-Hall, Inc, Englewood Cliffs, New Jersey, p 676

12. HAYWARD HE 1938 The structure of economic plants. MacMillan, New York, pp 451-484
13. HELLEBUST JA 1976 Osmoregulation. Annu Rev Plant Physiol 27:485-506
14. HIRAI M 1979 Sorbitol-6-phosphate dehydrogenase from loquat fruit. Plant Physiol 63:715-717
15. HIRAI M 1981 Purification and characteristics of sorbitol-6-phosphate dehydrogenase from loquat leaves. Plant Physiol 67:221-224
16. KENNEDY RA, D JOHNSON 1981 Changes in photosynthetic characteristics during leaf development in apple. Photosynth Res 2:213-223
17. KOW YW, DA SMITH, M GIBBS 1982 Oxidation of reduced pyridine nucleotide by a system using ascorbate and hydrogen peroxide from plants and algae. Plant Physiol 69:72-76
18. LAMBERS H, T BLACQUIERE, B (CEE) STUIVER 1981 Interactions between osmoregulation and the alternative respiratory pathway in *Plantago coronopus* as affected by salinity. Physiol Plant 51:63-68
19. LARCHER W 1969 The effect of environmental and physiological variables on the carbon dioxide gas exchange of trees. Photosynthetica 3:167-198
20. LATZKO E, M GIBBS 1981 Alkaline C_1-fructose-1,6-diphosphatase. *In* H Bergmeyer, ed, Methods of Enzymatic Analysis. Verlag Chemie Int, Deerfield Beach, Florida, pp 881-884
21. LEWIS DH, DC SMITH 1967 Sugar alcohols (polyols) in fungi and green plants. I. Distribution, physiology and metabolism. New Phytol 66:143-184
22. LOESCHER WH, GC MARLOW, RA KENNEDY 1982 Sorbitol metabolism and sink-source interconversions in developing apple leaves. Plant Physiol 70:335-339
23. LOESCHER WH, R REDGWELL, R BIELESKI 1982 Mannitol biosynthesis in higher plants: detection and characterization of a NADPH-dependent mannose 6-phosphate reductase in mannitol-synthesizing plants. Plant Physiol 69:S51

24. LURIE S, N PAZ, N STRUCH, BA BRAVADO 1979 Effect of leaf age on photosynthesis and photorespiration. In R Marcelle et al, eds, Photosynthesis and Plant Development. W Junk, The Hague, Netherlands, pp 31-38
25. McFADDEN BA, AR DENEND 1972 D-ribulose-1,5-diphosphate carboxylase from autotrophic microorganisms. J Bacteriol 110:633-640
26. MEHLER AH 1951 Studies on the reaction of illuminated chloroplasts. II. Stimulation and inhibition of the reaction with molecular oxygen. Arch Biochem Biophys 34:339-351
27. NEGM FB, WH LOESCHER 1979 Detection and characterization of sorbitol dehydrogenase from apple callus tissue. Plant Physiol 64:69-73
28. NEGM FB, WH LOESCHER 1981 Characterization of aldose-6-phosphate reductase (alditol-6-phosphate: NADP-1-oxidoreductase) from apple leaves. Plant Physiol 67:139-142
29. PARKER BL 1966 Translocation in Macrocystis. III. Composition of sieve-tube exudate and identification of the major ^{14}C-labelled products. J Physiol (London) 2:38-41
30. PIENIAZEK J, T HOLUBOWICZ, B MACHNIK, M KASPRZYK 1978 Apple stem callus frost tolerance and growth modification by adding sorbitol and some growth regulators to the medium. Acta Horticulture 81: 91-95
31. PLOUVIER V 1963 Distribution of aliphatic polyols and cyclitols. In T Swain, ed, Chemical Plant Taxonomy. Academic Press, New York, pp 313-336
32. POPP M 1984 Chemical composition of Australian mangroves. II. Low molecular weight carbohydrates. Z Pflanzenphysiol 113:411-421
33. POWLES SB 1984 Photoinhibition of photosynthesis induced by visible light. Annu Rev Plant Physiol 35:15-44
34. RADNER RJ, B KOK, O OLLINGER 1978 Kinetics and apparent K_m of oxygen cycle under conditions of limiting carbon dioxide fixation. Plant Physiol 61:915-917
35. RAESE JT, MW WILLIAMS, HD BILLINGSLEY 1977 Sorbitol and other carbohydrates in dormant apple shoots as influenced by controlled temperatures. Cryobiology 14:373-378

36. REDGWELL RJ 1980 Fractionation of plant extracts using ion-exchange sephadex. Anal Biochem 107:44-50
37. REDGWELL RJ, RL BIELESKI 1978 Sorbitol-1-phosphate and sorbitol-6-phosphate in apricot leaves. Phytochemistry 17:407-409
38. ROBINSON JM 1984 Photosynthetic carbon metabolism in leaves and isolated chloroplasts from spinach plants grown under short and intermediate photosynthetic periods. Plant Physiol 75:397-409
39. RUFTY TW JR, PS KERR, SC HUBER 1983 Characterization of diurnal changes in activities of enzymes involved in sucrose biosynthesis. Plant Physiol 73:428-433
40. RUMPHO ME, GE EDWARDS, WH LOESCHER 1983 A pathway for photosynthetic carbon flow to mannitol in celery leaves. Plant Physiol 73:869-873
41. SAKAI A 1966 Seasonal variations in the amounts of polyhydric alcohol and sugar in fruit trees. J Hort Sci 41:207-213
42. SAYRE RT, RA KENNEDY, DJ PRINGNITZ 1979 Photosynthetic enzyme activities and localization in *Mollugo verticillata* populations differing in the levels of C_3 and C_4 cycle operation. Plant Physiol 64:293-299
43. STACEY BE 1974 Plant polyols. In JB Pridham, ed, Plant Carbohydrate Biochemistry. Academic Press, New York, pp 47-59
44. STITT M, HW HELDT 1981 Physiological rates of starch breakdown in isolated spinach chloroplasts. Plant Physiol 68:755-761
45. STITT M, T AP REES 1980 Carbohydrate breakdown by chloroplasts of *Pisum sativum*. Biochim Biophys Acta 627:131-143
46. TOLBERT NE 1980 Photorespiration. In DD Davis, ed, The Biochemistry of Plants, Vol 2. Academic Press, New York, pp 487-523
47. TOUSTER O, DRD SHAW 1962 Biochemistry of the acyclic polyols. Physiol-Rev 42:181-225
48. TRIP P, G KROTKOV, CD NELSON 1963 Biosynthesis of mannitol-C^{14} from $C^{14}O_2$ by detached leaves of white ash and lilac. Can J Bot 41:1005-1010
49. TRIP P, G KROTKOV, CD NELSON 1964 Metabolism of mannitol in higher plants. Am J Bot 51:828-835
50. WEBB KL, JWA BURLEY 1962 Sorbitol translocation in apple. Science 137:766

51. WILLENBRINK J, BP KREMER, V SCHMITZ, M WEIDNER 1979 CO_2-Fixierung und Stofftransport in benthischen marinen Algen. Ber Dtsch Bot Ges 92:157-167
52. WILLIAMS LE, RA KENNEDY 1978 Photosynthetic carbon metabolism during leaf ontogeny in Zea mays L.: enzyme studies. Planta 142:269-274
53. YAMAKI S 1980 A sorbitol oxidase that converts sorbitol to glucose in apple leaf. Plant Cell Physiol 21:591-599
54. YAMAKI S 1984 $NADP^+$-dependent sorbitol dehydrogenase found in apple leaves. Plant Cell Physiol 25:1323-1327
55. YANCEY PH, ME CLARK, SC HAND, RD BOWLES, GN SOMERO 1982 Living with water stress: evolution of osmolyte systems. Science 217:1214-1222
56. ZEIGLER H 1975 Nature of transported substances. In MH Zimmerman, JA Milburn, eds, Encyclopedia of Plant Physiology, Vol 1, New Series. Springer-Verlag, New York, pp 59-136
57. ZIMMERMAN MH, H ZEIGLER 1975 List of sugars and sugar alcohols in sieve-tube exudates. In MH Zimmerman, JA Milburn, eds, Encyclopedia of Plant Physiology, Vol 1, New Series. Springer-Verlag, New York, pp 480-503

Interrelationships Between Photosynthetic Carbon and Nitrogen Metabolism in Mature Soybean Leaves and Isolated Leaf Mesophyll Cells

J. MICHAEL ROBINSON and CHRIS BAYSDORFER

Increased metabolic "sink" strength in roots and above-ground organs brought on by demand of growth and development processes for new photosynthate, often results in an increase in net photosynthetic CO_2 assimilation rate (22, 32, 36). In reality, increased inorganic nitrogen assimilation may express a metabolic sink for newly-generated carbohydrates (3), even in the sink leaves, resulting in increased photosynthetic rate in source leaves. That challenge of mature, intact plants and isolated leaf mesophyll cells by NO_3^- or NH_4^+ can cause an increase in photosynthetic CO_2 fixation rate which has been observed to be quite variable (9, 23, 26, 27, 33, 38, 39). The conditions under which NO_3^-, NO_2^-, and NH_4^+ may stimulate, or repress, photosynthetic rate requires much more examination, especially in higher plant leaves.

In leaves, the interrelationship between carbon and nitrogen assimilation is far more complex than it is in roots, and the picture of how these processes interact and compete for photolytically-derived reductant is not clear. Foliar reductive inorganic nitrogen assimilation, from NO_3^- to α-amino nitrogen is almost totally dependent upon light (through reduced ferredoxin) (1, 3, 8, 30, 31). Nitrate reduction to nitrite in the mesophyll leaf cytoplasm (1) is dependent upon reduced pyridine nucleotide derived from reducing equivalents originating from the photolysis of H_2O, and transported out of the chloroplast as triose-phosphate (triose-P) and malate across the plastid envelope via triose-P-orthophosphate and C_4 dicarboxylate shuttle systems (1, 5, 13, 14, 17). The resulting transport into the chloroplast, is reduced to α-amino nitrogen employing reductant and ATP derived from H_2O photolysis (1, 2, 8). Additionally, due to photorespiratory NH_4^+ release, ammonia is also reassimilated in the chloroplast (by glutamine synthetase) (37, 40). It would seem that the demands for reductant and ATP for foliar nitrogen assimilation are met almost totally by the chloroplast, and that carbon assimilation demands for reductant and ATP could take precedence over those for nitrogen assimilation, but this remains an important issue (8). Rather than repression of carbon assimilation by inorganic nitrogen, Woo and Canvin (38), as well as Paul et al. (26), have demonstrated that challenge of photosynthetic leaf cells with ammonia stimulates CO_2 photoassimilation. Also, Bassham and his associates (7, 19, 20, 26, 27) have repeatedly shown that NH_4^+ stimulates the anaplerotic pathway in photosynthetic cells including a stimulation of enzymes, most prominently, pyruvate kinase and phosphoenolpyruvate carboxylase (7, 26). Also, Plaut et al. (28) observed that photosynthetic intermediates such as dihydroxyacetone phosphate also stimulated NO_2^- photoreduction in a reconstituted spinach leaf plastid system. Robinson and Snyder (33) observed that soybeans, fed levels of NO_3^- plus NH_4^+ which were sufficient or above to promote growth, displayed at least a 2-fold increase in net CO_2 photoassimilation.

This study has the purpose of examining, in mature soybean leaves and in intact mesophyll cells isolated from those leaves, the interrelationships between carbon and nitrogen photoassimilation. Specifically, the influence that NO_3^-, NO_2^-, and NH_4^+ exerts upon net CO_2 photoassimilation rate in leaves and isolated mesophyll leaf cells will be discussed in relationship to (1) competition between carbon and nitrogen photoassimilation for photolytically-derived reductant (reduced ferredoxin) or (2) how foliar nitrogen assimilation can stimulate photosynthetic carbon metabolism. Additionally, consideration will be given to the influence that inorganic nitrogen assimilation exerts upon foliar photosynthate partitioning between (a) the pentose phosphate reductive cycle, (b) the anaplerotic pathway*, and (c) the synthesis of foliar starch, sucrose, and total amino acids.

MATERIALS AND METHODS

<u>Plant Material</u>. Nonnodulated, non-N_2 fixing soybean plants (cv Williams and cv Amsoy) were propagated in vermiculite in 6" plastic pots held continuously in the growth chamber. Soybean plants were fed a modified Hoagland's solution containing increasing levels of NO_3^- plus NH_4^+ in a ratio of 5 to 1 (mM basis). Composition of the nutrient solutions with which plants were irrigated with optimal NO_3^- plus NH_4^+ levels has been previously reported (32).

<u>Net Photosynthesis in Trifoliates</u>. Net photosynthetic carbon assimilation (PS) was estimated employing infrared gas analysis methods previously described (32); analyses were performed on trifoliolates which remained attached to the parent plants. Leaves were then removed from the plants and leaf area, as well as fresh weight, was determined; the leaves were immersed in liquid N_2, and then subjected to lyophilization (32). Chl was determined on this freeze-dried leaf tissue (32). Total leaf nitrogen was measured using standard Micro-Kjeldahl digestion and autoanalyzer techniques.

<u>Isolated Mesophyll Cells and Reconstituted Chloroplast Systems</u>. Mesophyll cells were isolated mechanically in the absence of inorganic nitrogen (25) from trifoliolates, #3-6 number acropetally, for 25 to 30 d-old soybean (Amsoy) grown as above, irrigated with the nutrient solution containing 9 mM total nitrogen. However, to increase the rapidity of cell isolation, centrifugation (50g/1.5 min) was employed to remove cells from the homogenization medium and the washing medium. This isolation of cells required only 20 to 30 min.

---

*In the context of this discussion, the anaplerotic pathway is described as the flow of PGA and triose-P out of the chloroplast, into the cytoplasm where these intermediates are metabolized to both pyruvate (Pyr) and phosphoenolpyruvate (PEP), which are then further metabolized to organic acids (<u>e.g</u>. malate and oxalacetate) as well as to amino acids (<u>e.g</u>. alanine). See Figure 8.

$^{14}CO_2$ photoassimilation was measured employing methods described by Oliver et al. (25). Cells were assayed in 3.5 to 4.0 ml reaction mixtures containing 50 mM HEPES, pH 7.8; 0.33 sorbitol; 2 mM DiKEDTA; 1 mM $CaCl_2$; 1 mM $MnCl_2$; 0.5 mM K_2HPO_4-KH_2PO_4; 2 mM DTT; 0.5 to 5.0 mM "CO_2" (CO_2 dissolved plus HCO_3^-) with 10 to 40 uCi-$^{14}CO_2$; and 0.25 to 2.5 mM $NaNO_3$, $NaNO_2$, or NH_4Cl, where indicated. Nitrite photoreduction was measured employing previously described techniques (22, 30, 35). Where applicable, $^{14}CO_2$ photoassimilation and NO_2^- photoreduction were simultaneously monitored in the same reaction mixtures. All experiments were carried out at a radient flux density of 1000 µmol photon m^{-2} s^{-1} and at a temperature of 25°C.

Procedures involving measurements of simultaneous NADP and NO_2^- photoreduction in reconstituted spinach leaf chloroplast systems have been reported previously by Baysdorfer and Robinson (8).

Foliar Photosynthate and Metabolites. Sucrose and starch were estimated in extracts from the freeze-dried trifoliolate tissue (20 mg), as previously described (32). PGA, Pyr, and triose-P were measured enzymatically in perchlorate precipitated extracts from lyophilized tissues, as described previously (10, 32). Extracts were prepared from lyophilized foliar tissue (deveined) by a heat step with a zinc acetate precipitation procedure described by Scholl et al. (35). Amino nitrogen (free amino acids) were estimated, employing the ninhydrin procedures described by Yemm and Cocking (41).

Enzymatic Activities. Ribulose-1,5-bisphosphate carboxylase/oxygenase (Rubisco), fructose-1,6-bisphosphate (Fru-1,6-P_2),1-phosphatase (FbPase, at pH 8.7, chloroplast enzyme), PEP carboxylase (PEPCase), and Pyr kinase (PKase) activities were estimated in extracts prepared from the lyophilized leaves (described above). Samples (100 mg) were extracted with 10 ml of medium containing 50 mM HEPES, pH 7.6; 2 mM DTT; 1 mM DiKEDTA; and 0.5% (w/v) PVP-40 in Potter-Elvehjem homogenizers. This homogenate was centrifuged (4°C) at 32,837g/20 min; the resulting pellet was re-extracted with 3 ml of homogenization medium. Protein was estimated by the Bradford method (9). Rubisco was estimated using the combined methods of Bahr and Jensen (4) and Lilley and Walker (21); FbPase was estimated employing the method of Kelly et al. (16); PEPCase was estimated employing the method of Lane et al. (18); and PKase was estimated employing the methods noted in Sarkissian and Fowler (34).

PHOTOSYNTHETIC RATE AND INORGANIC NITROGEN ASSIMILATION IN MATURE LEAVES AND ISOLATED LEAF CELLS

Foliar Net CO_2 Photoassimilation and Increasing NO_3^- and NH_4^+. Responses of foliar CO_2 photoassimilation were examined in three sets of simultaneously-propagated soybean plants (cv Williams) supplied daily with nutrient media containing NO_3^- plus NH_4^+ (Fig. 1). Each set was propagated with a different N level. The N levels were, with respect to growth (Fig. 1A), either slightly suboptimal (3 mM) (70% of normal growth), threshold optimal (9 mM), or superoptimal (18 mM). The ratio of NO_3^- to NH_4^+ in the nutrient media was maintained at 5 to 1, which resulted in increases in both NO_3^- and NH_4^+ as N level was increased. Gamborg and his associates (6, 12) observed that while growth of root cells cultured from several species was not stimulated by NH_4^+ supplied along with NO_3^- in the culture medium, maximal growth of cultured soybean root cortical cells required that NH_4^+ or glutamine be supplied along with NO_3^-. Thus, both NO_3^- and NH_4^+ were included in the nutrient solutions used to irrigate the soybean plants.

Figure 1B indicated that at 9 mM N, leaf growth was nearly maximal, and based upon the foliar Chl measurements, the chloroplasts were fully developed (Fig. 1B). Since leaf and total plant growth were increased by increasing N from 3 to 9 mM, it was not surprising to find that there was an increase in net photosynthesis (PS) in this situation (Fig. 1, A and D). Furthermore, doubling the N level from 9 to 18 mM produced ca. 5% increase in leaf growth, but there was ca. 1.5- to 1.6-fold additional increase in net PS in the trifoliolates measured (Fig. 1, A and D).

Marques et al. (23) reported that NO_3^-, compared with NH_4^+ nutrition, will stimulate the Calvin cycle assimilation of bean plants. The work shown in Figure 1, as well as that reported by other laboratories (23, 26, 38), raises several questions concerning the influence that nitrogen assimilation exerts upon photosynthetic carbon assimilation. The acquisition and assimilation of inorganic nitrogen in root tissue is an active

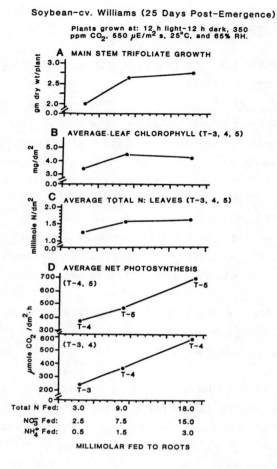

FIG. 1. Influence of increasing NO_3^- plus NH_4^+ fed to soybean plants propagated in vermiculite pots in the growth chamber. A, Main stem trifoliolate growth; B, Average leaf Chl; C, Average foliar total nitrogen; and D, Average net photosynthesis. In D, in order to compare leaves of similar sizes, trifoliolate 3, (T-3) (acropetally numbered) of the lowest N level plant was compared with T-4 of the optimal and super-optimal N plants; T-4 of the lowest N level plants was compared with T-5 of the higher N level fed plants.

metabolic sink (3, 42), and the increased inorganic N could cause the stimulation of a higher tissue sink demand for photosynthate from the foliage. Indeed, it is clear that increased tissue sink demand can increase net PS rate (24, 32, 36), and research over the span of three decades has made it clear that nitrogen assimilation in nonphotosynthetic as well as photosynthetic tissues depends heavily on newly-synthesized photosynthate (3, 42). However, increased sink demand still influences photosynthetic rates even after 1 or more days of acclimation of the source leaves (24, 32, 36). It is possible that inorganic nitrogen, transported to the source leaves, actually influences photosynthetic rate at the point in which inorganic N influxes into mesophyll leaf cells. In this case, there is no gradual sink demand acclimation required prior to an increase in source leaf net PS. This question was examined by employing soybean leaf mesophyll cells isolated from the mature leaves of plants grown with optimal N.

Net PS and Inorganic N in Isolated Soybean Leaf Mesophyll Cells. Soybean mesophyll cells were isolated from mature foliage of plants propagated in a nitrogen-sufficient nutrient culture (10 mM N). When the "CO_2" was saturating with respect to net PS, it was found that not only NO_3^- (Fig. 2A, B), but also NH_4^+ (Fig. 2C) and NO_2^- (Fig. 2D) stimulated net CO_2 photoassimilation almost 2-fold in some cases. However, when "CO_2" levels were rate-limiting, only slight or no stimulation of net PS was brought on by NO_3^-, NH_4^+, or NO_2^- (Fig. 2A, D, compare high with low "CO_2"). Indeed, Woo and Canvin found that NH_4^+ produced little or no stimulation of net PS in isolated spinach leaf cells when the "CO_2" level was rate-limiting (38), and that neither NO_3^- nor NO_2^- had any effect on net PS rate (39).

Although net PS of isolated spinach leaf cells did not respond to NH_4^+ below 1 mM (38), it was found that in soybean cells, NH_4^+, NO_3^-, and NO_2^- were effective in stimulating net PS at levels as low as 250 μm (Fig. 2C and data not shown). This indicated that net PS in soybean foliage responded to levels of inorganic N which are often found at these concentrations (e.g. NO_2^- levels of 120 to 190 μm; JM Robinson, unpublished).

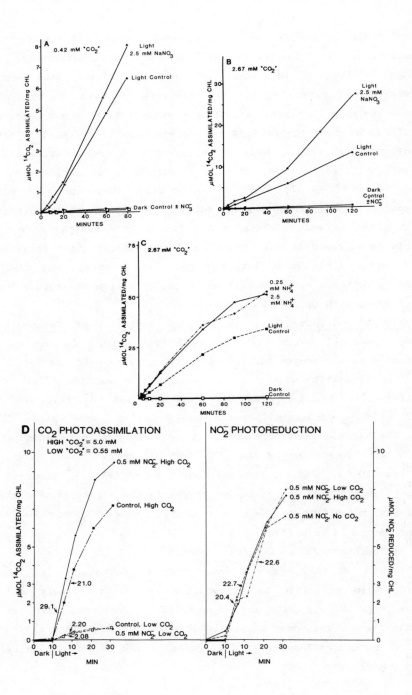

FIG. 2. Effect upon isolated soybean-leaf mesophyll net $^{14}CO_2$ photoassimilation of NO_3^- (A, B), NH_4^+ (C), and NO_2^- (D). In A, $NaNO_3$ (2.5 mM) was fed to cells photosynthesizing in the presence of "CO_2", rate-limiting (0.42 mM) with respect to CO_2 fixation. B, NO_3^- level as in A, but "CO_2" was rate-saturating (2.67 mM) with respect to cell CO_2 assimilation. C, NH_4Cl was fed at 0.25 and 2.5 mM, and "CO_2" (2.67 mM) was rate saturating. D, photoassimilation and nitrite photoreduction were monitored simultaneously where applicable. $NaNO_2$ was supplied at 0.5 mM to mesophyll cell isolates in which there was either no "CO_2", 0.5, or 5.0 mM supplied. See "Methods" sections for reaction mixture composition. Reaction mixture volume for A, B, and C was 4.0, and for D, 3.5 ml. Whole cell Chl in the reaction mixtures was, for A and B, 30.71 µg; for C, 64.3 µg; and for D, 29.7 µg. Rates of $^{14}CO_2$ photofixation (µmol mg^{-1} Chl h^{-1}) at the most maximal values (linear) were: for A, control and with NO_3^-, respectively, 5.3 and 4.8; for B, control and with NO_3^-, respectively, 7.5 and 19.5; for C, control and with NH_4^+, respectively, 20 and 38. In D, most maximal rates of $^{14}CO_2$ fixation were, at saturating "CO_2" and without NO_2^-, 29.1, and with NO_2^-, 21.0; at rate-limiting "CO_2." CO_2 fixation rates were, without and with NO_2^-, respectively, 2.2 and 2.1. Maximal NO_2^- photoreduction rates were ca. 22 µmol mg.$^{-1}$ Chl h^{-1} regardless of the level of "CO_2" for CO_2 photoassimilation.

It must be concluded from these studies (Fig. 1, 2), that the stimulation of net PS, in the case of soybean foliage, may occur when NO_3^-, NO_2^-, and NH_4^+ directly enter the soybean leaf mesophyll cell. Increased net PS rate may not necessarily be entirely attributable to increased sink demand for photosynthate required to drive N assimilation in nonphotosynthetic tissues. This does not rule out the possibility that increasingly higher levels of N caused, in mature leaves, increases in the activities of foliar carbon metabolism enzymes. This point will be discussed in the Photosynthate Partitioning section of this report.

<u>Do Foliar CO_2 and NO_2^- Photoassimilation Compete for Reductant?</u>
Competition between the reductant and energy demands of CO_2 photoassimilation and those of NO_2^- photoreduction in higher plant chloroplasts has been a long-term controversy (8). Since it was clear that inorganic N would stimulate net CO_2 photoassimilation, it was important to understand whether or not there had been an increase or decrease in the rate of light-dependent assimilation of NO_2^- in the chloroplast during concurrent CO_2 assimilation. This would answer the question pertaining to whether or not CO_2 assimilation and N assimilation were forced to compete <u>in situ</u> for reduced ferredoxin.

The results shown in Figure 2D demonstrated that while NO_2^- stimulated CO_2 assimilation in soybean leaf mesophyll cells, there was no effect upon the light-dependent reduction of NO_2^- by CO_2 assimilation. Nitrite photoreduction displayed a rate of ca. 22 $\mu mol\ mg^{-1}\ Chl\ h^{-1}$ regardless of the level of "CO_2" present for $^{14}CO_2$ fixation. Even when reductant demand by CO_2 assimilation was maximal, <u>e.g.</u> saturating "CO_2", there was no repression of NO_2^- reduction (Fig. 2D). Further, CO_2 photoassimilation displayed highest rates (30 $\mu mol\ mg^{-1}\ Chl\ h^{-1}$) in the presence of NO_2^- reduction. Thus, NO_2^- reduction did not repress CO_2 assimilation. In isolated intact spinach leaf plastids, as in soybean leaf cell preparations, NO_2^- photoreduction was observed to have rates of 8 to 25 $\mu mol\ mg^{-1}\ Chl\ h^{-1}$, regardless of the reductive demands for CO_2 assimilation which displayed rates of 40 to 60 $\mu mol\ mg^{-1}\ Chl\ h^{-1}$ (J. M. Robinson, unpublished). This is interpreted to mean that NO_2^- reduction does not have to compete for, nor is it repressed by, reductive demands for reduced ferredoxin and

NADP reduction brought on by carbon assimilation in the chloroplast, as long as the NADPH requirements are not extremely high.

That there can be rather strong competition for reduced ferredoxin between NADP and NO_2^- photoreduction is shown in Figure 3. In a reconstituted spinach leaf chloroplast system (stroma, lamellae, and saturating levels of purified ferredoxin, NADP, and NO_2^-) photoreduction of NADP at very high rates (310 µmol mg^{-1} Chl h^{-1}), results in repression of concurrent NO_2^- reduction until almost all the NADP has been reduced (Fig. 3 and ref. 8). In this case, it seems clear that ferredoxin-NADP reductase has preference over nitrite reductase for reduced ferredoxin. It may be that the K_m of nitrite reductase for reduced ferredoxin in the reconstituted systems is higher than that of ferredoxin-NADP reductase. This would result in the preferential flow of electrons to NADP at low reduced ferredoxin levels. Also, there is the possibility that the reconstituted system is not correctly reassembled, so that nitrite reductase is not correctly juxtaposed to the ferredoxin reducing site.

Alternatively, if the reconstituted system (Fig. 3B) actually reflects the true in situ situation, then, when rates of CO_2 photoassimilation were 200 to 400 µmol mg^{-1} Chl h^{-1}, the reductive demands of carbon assimilation would repress those of nitrogen assimilation. At this point, it must be concluded that, in vivo, in the intact soybean cell and intact spinach leaf plastid, when CO_2 assimilation rates are moderate, there is sufficient reduced ferredoxin to supply both CO_2 photoassimilation and NO_2^- reduction. Apparently there is still enough additional ferredoxin to permit an increase in CO_2 fixation during NO_2^- assimilation, without a decrease in NO_2^- reduction (Fig. 2D). Conceivably, in the intact system, there may be photosynthetic electron transport chains, other than those coupled to ferredoxin-NADP reductase, which serve the processes of nitrogen photoassimilation such that CO_2 and N photoassimilatory processes never compete for reductant.

PHOTOSYNTHATE PARTITIONING AND INORGANIC N

Enzyme Sites Stimulated by Increasing Inorganic N. The question remained as to exactly why in soybean the net PS increased in response to supplied inorganic N; i.e., was there an increase in the activities of

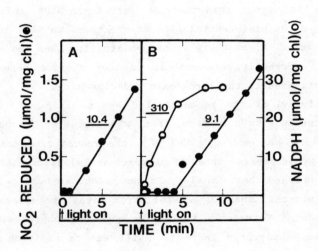

FIG. 3. Nitrite photoreduction in a reconstituted spinach chloroplast system (lamella plus stroma) in the absence (A) and presence (B) of concurrent NADP reduction. Reaction mixtures contained 1 mM ADP, 1 mM GSH, 1 mM GSSG, 3 mM Pi, 10 uM ferredoxin, 225 ug stromal protein, and 43 ug Chl (lamellae). Nitrite and NADP reduction were initiated with the addition of 0.4 mM NO_2^- and 1.0 mM NADP (initial concentrations in the reaction mixtures). The experiments were carried out at a radiant flux density of 3000 μmol Photon m^{-2} s^{-1}. Underlined numbers are reaction rates in units of μmol mg^{-1} Chl h^{-1}.

Calvin cycle or other carboxylating enzymes? What metabolic changes occurred in the soybean foliar carbon and nitrogen pathways to accommodate the assimilation of added inorganic nitrogen?

Fowler and his associates (11, 15, 34) working with nonphotosynthetic tissues (e.g. roots) have made it clear that when NO_3^- is being acquired and assimilated, the pentose phosphate-pathway enzymes--including glucose-6-P dehydrogenase, 6-phosphogluconate dehydrogenase, and transketolase--are stimulated in order to provide NADPH for reductant (11, 15, 34). Additionally, it was clear that root PKase is also stimulated by inorganic nitrogen (15, 34).

Bassham and his associates (7, 19, 20, 26, 27) have found that in leaf discs and in isolated mesophyll cells, NH_4^+ stimulated the anaplerotic pathway (metabolism from triose-P and PGA to PEP, Pyr, organic acids, and amino acids) (Fig. 8). This may involve activation of a foliar PKase isoform (7). Paul et al. (26) observed that NH_4^+ stimulated PEPCase activity. On the other hand, Marques et al. (23) observed, in young bean plants, that the Calvin cycle was stimulated by NO_3^-.

In soybean leaves, it appears that both the Calvin cycle enzymes as well as those of the anaplerotic pathway are stimulated by increasing levels of NO_3^- and/or NH_4^+. Increase of inorganic N level from 3 to 9 mM resulted in significant increases in the total activity of Rubisco, FbPase (pH 8.7, plastid enzyme), PKase, and PEPCase, which appeared to be correlated with the increases in the net PS rate (Fig. 4A, B). Importantly, the additional increase of N from 9 to 18 mM resulted in still additional increased activity of foliar PEPCase, PKase, and FbPase. Although increases in N level appeared not to increase the foliar specific activity of Rubisco, the other three enzymes displayed significant specific activity increases (Fig. 4C).

Enzyme analyses suggest that increased N level (Fig. 1) caused an increase in Calvin cycle activity, as well as an increase in PEPCase and PKase. The step increase in net CO_2 fixation may result, in part, from an increase in activity of both Rubisco and PEPCase (Fig. 4A, B, C). These data are consistent with the hypothesis that the stimulation of net PS by increasing N was the result of both an increase in Calvin cycle activity and an increase in the anaplerotic pathway (see Fig. 8).

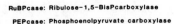
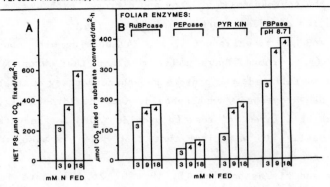
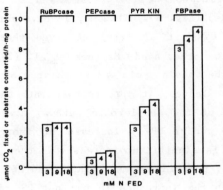

FIG. 4. Influence of increasing level of N (NO_3^- plus NH_4^+) upon foliar enzymes associated with the Calvin cycle and the anaplerotic carbon flow pathway. Trifoliolates (e.g. T-3 from plants fed lowest inorganic N levels compared with T-4 of plants fed optimal or super-optimal levels of N) from the Figure 1 study were monitored for net photosynthesis (A), and subsequently these leaves were immersed in liquid N_2 and then lyophilized to dryness. Freeze-dried trifoliates (from the photosynthesis study) were extracted and the activities of Rubisco, FbPase (pH 8.7), PEPCase, and PKase were estimated and expressed on a leaf-area basis (B), and on a soluble leaf-protein basis (C).

Metabolite Flow. Similar to the enzyme analyses, metabolite analyses in the trifoliolates of the plants shown in the Figure 1 experiment reflected that increasing N level influenced the intact leaf primary carbon and nitrogen metabolism. When compared with plants fed inorganic N, the other N-level plants displayed step-wise diminution of foliar level and synthesis of starch (Fig. 5C, D), as well as concomitant decreases in the pool of PGA (Fig. 6A) in all leaves tested. The rather sharp diminution of both PGA and starch synthesis as a function of increased N level, was interpreted to mean that the reduction of starch synthesis was related to the lowering of PGA level. Since foliar Pi levels may either decrease slightly or remain approximately constant when net PS rate increases (32), then the PGA/Pi ratio would decrease, which would result in inhibition of ADP-glucose pyrophosphorylase (29, 32). The end result would be a limitation on the rate and level of accumulation of starch (29). It was recently observed that changes in the foliar PGA/Pi ratio were correlated with the rate of foliar starch synthesis (32). Certainly, it is clear that increased nitrogen influx places a constraint upon starch synthesis (Fig. 5).

The influence that increased N exerts upon decreasing foliar starch levels seems to reflect that carbon metabolites, rather than being sequestered as starch, are being exported out of the chloroplast into the rest of the leaf cell for the production of amino acid carbon skeletons as more organic N becomes available (Fig. 5A). It seems likely that one of the major precursors of amino acids in this situation is PGA and triose-P. The decrease of PGA with increased N has already been discussed. In plants fed 9 mM N compared to those fed 3 mM, there was a significant increase in triose-P (Fig. 6B). The pool size sharply declined when the N level was increased further to 18 mM. A similar pattern was observed in the case of pyruvate (Fig. 6C). A significant increase in the free foliar amino acids in the 9 mM N plants was also observed (Figs. 5A, 7A, B). Indeed, the amino acid accumulation rate from morning to afternoon in the plants fed 9 mM N was 2.4-fold higher than in the plants fed 3 mM N (Fig. 7B), even though there was only a 1.3-fold increase in leaf growth, respectively. This meant that an increase of N from 3 to 9 mM, resulted in an increase in amino acid accumulation (in the light) which was considerably in excess of that required to satisfy the needs for leaf growth.

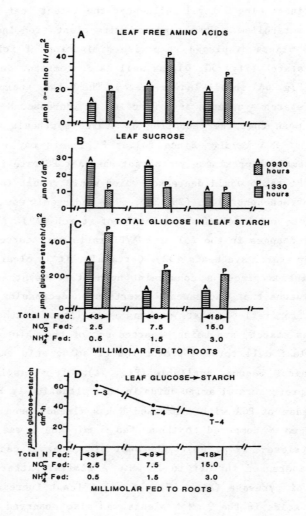

FIG. 5. Partitioning of photosynthate as a function of inorganic N level. From the Figure 1 experiment. In T-3, compared with T-4, foliar free amino acid levels at 0930 and 1330 h (A) are compared with sucrose levels (B), leaf starch (glucose) levels (C), and starch synthesis rates (D).

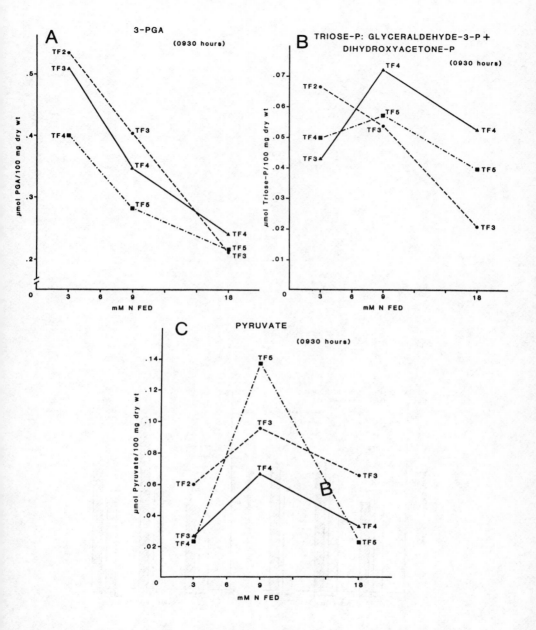

FIG. 6. Levels of foliar PGA (A), triose-P (B), and pyruvate (C) as a function of the level of inorganic N fed to soybean plants. From the Figure 1 study. Samples are from the 0930 h experiment.

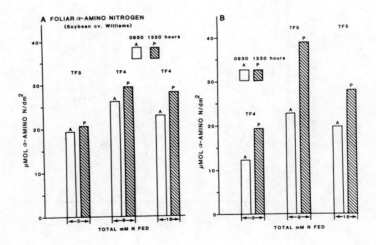

FIG. 7. Accumulation of foliar amino acids in tissues. From the Figure 1 experiment, see above.

It is also suggested that the flow of carbon from triose-P through the anaplerotic pathway to amino acid synthesis was accelerated with the increase in N. Ammonia, fed to either photosynthesizing alfalfa leaf discs (27), or to isolated, intact mesophyll cells (19, 20, 26) has been observed to enhance the metabolism of PEP through Pyr to alanine through the anaplerotic pathway (Fig. 8). Bassham and his associates (19, 20, 26, 27) also demonstrated that in cells and leaves of C-3 plants, there is an enhanced conversion of amino acids to amides, e.g. glutamate to glutamine, upon challenge of photosynthetic cells or tissues with NH_4^+.

One of the important questions centers around whether or not the newly synthesized PGA is exported from the chloroplast directly to the formation of Pyr, or alternatively, whether or not sucrose is the precursor of Pyr through glycolytic conversion of sucrose to triose-P, and then to Pyr through the anaplerotic pathway. Bamberger et al. (5) observed that the export of triose-P out of the chloroplast via a shuttle system and involving the nonreversible triose-P dehydrogenase, resulted in an increase in the rate of CO_2 fixation in isolated intact spinach leaf chloroplasts. In that system, sufficient carbon metabolites were maintained in the Calvin cycle to support the regeneration of Ru-1,5-P_2, and for export of triose-P into the cytoplasm. This study is reminiscent of the metabolic patterns brought on when inorganic N begins to enter the soybean leaf cell. Certainly, in photosynthetic cells and tissues, incorporation of NH_4^+ may occur at the expense of sucrose (19, 26, 27), but in most observations, it was difficult to tell whether or not metabolite flow had been redirected away from sucrose, or was a direct drain upon the sucrose pool. It must be concluded, based upon the influence that normal N levels (growth sufficient) exerted upon PGA, triose-P levels, as well as upon photosynthetic rate (Figs. 1, 2, 4, 5-7), that when an influx of NO_3^-, NO_2^-, and/or NH_4^+ occurs, there is an increase in net PS. Then sufficient PGA and triose-P is formed so that not only is there adequate metabolite flow to support sucrose synthesis, but there is additional adequate flow to support amino acid metabolism. In other words, the ultimate precursors of the amino acids appear to be the PGA and triose-P formed in the chloroplast, and the enhanced export of these compounds into the cytoplasm, in response to the demands of the ana-

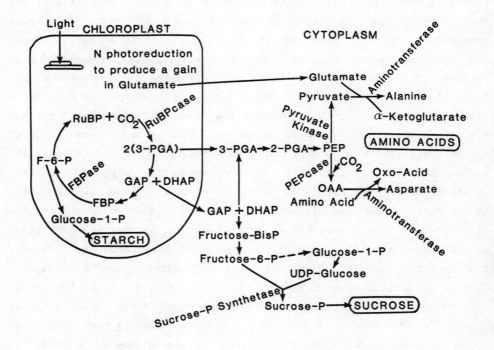

FIG. 8. A metabolite flow diagram for the chloroplast and cytoplasm emphasizing photosynthate flow to starch, to sucrose, and through the anaplerotic pathway to the amino acid pools.

plerotic pathway, triggers an increase in the PS rate. The mechanism associated with this PS increase remains as a future goal for study. At very high N levels (18 mM), apparently a large demand for carbon photosynthate, ostensibly throughout the entire plant, forces the mobilization of even sucrose reserves (Fig. 5B), but this point is not clear.

Finally, this study emphasizes the very important influence that whole plant and foliar nitrogen assimilation exerts upon the rate of PS, as well as upon the direction of metabolite flow during photosynthetic carbon metabolism. It is clear that models of regulation of carbon photosynthate partitioning in crop plant leaf cells must include a thorough consideration of the influence that foliar nitrogen assimilatory events initiate.

ACKNOWLEDGMENTS--The authors thank Mr. D. R. Lee for net PS measurements, Mr. W. F. Stracke for technical assistance, Dr. F. W. Snyder for plant growth-nutrient culture design, Ms. M. F. O'Brien for typing the manuscript, and J. M. R. thanks Dr. Z. Plaut for a very stimulating discussion during manuscript preparation.

LITERATURE CITED

1. ANDERSON JW 1981 Light-energy-dependent processes other than CO_2 assimilation. In MD Hatch, NK Boardman, eds, The Biochemistry of Plants, Vol 8, Photosynthesis. Academic Press, New York, pp 473-500
2. ANDERSON JW, DA WALKER 1983 Oxygen evolution by a reconstituted spinach chloroplast system in the presence of L-glutamine and 2-oxoglutarate. Planta 159:77-83
3. ASLAM M, RC HUFFAKER, DW RAINS, KP RAO 1979 Influence of light and ambient carbon dioxide concentration on nitrate assimilation by intact barley seedlings. Plant Physiol 63:1205-1209
4. BAHR JT, FG JENSEN 1978 Activation of ribulose bisphosphate carboxylase in intact chloroplasts by CO_2 and light. Arch Biochem Biophys 185:39-48
5. BAMBERGER ES, BA EHRLICH, M GIBBS 1975 The glyceraldehyde-3-phosphate and glycerate-3-phosphate shuttle and carbon dioxide assimilation in intact spinach chloroplasts. Plant Physiol 55:1023-1030

6. BAYLEY JM, J KING, OL GAMBORG 1972 The effect of the source of inorganic nitrogen on growth and enzymes of nitrogen assimilation in soybean and wheat cells in suspension cultures. Planta 105:15-24
7. BAYSDORFER C, JA BASSHAM 1984 Spinach pyruvate kinase isoforms: Partial purification and regulatory properties. Plant Physiol 74: 374-479
8. BAYSDORFER C, JM ROBINSON 1985 Metabolic interactions between spinach leaf nitrite reductase and ferredoxin-NADP reductase: competition for reduced ferredoxin. Plant Physiol 77:318-320
9. BRADFORD M 1976 A rapid and sensitive method for the quantitation of microgram quantities of protein utilizing the principle of protein-dye binding. Anal Biochem 72:248-254
10. CZOK R, W LAMPRECHT 1974 Pyruvate, phosphoenolpyruvate, and D-glycerate-2-phosphate. In HU Bergmeyer, K Gawehn, ed, Methods of Enzymatic Analysis. Verlag Chemie and Academic Press, NY, Vol. 3, pp 1446-1451
11. EMES MJ, MW FOWLER 1983 The supply of reducing power for nitrite reduction in plastids of seedling pea roots (Pisum sativum L.). Planta 158:97-102
12. GAMBORG OL 1970 The effects of amino acids and ammonium on the growth of plant cells in suspension culture. Plant Physiol 45: 372-375
13. HEBER U, HW HELDT 1981 The chloroplast envelope: structure, function, and role in leaf metabolism. Annu Rev Plant Physiol 32:139-168
14. HOUSE CM, JW ANDERSON 1980 Light dependent reduction of nitrate by pea chloroplasts in the presence of nitrate reductase and C4-dicarboxylic acids. Phytochem 19:1925-1930
15. JESSUP W, MW FOWLER 1977 Interrelationships between carbohydrate metabolism and nitrogen assimilation in cultured plant cells, III. Effect of the nitrogen source on the pattern of carbohydrate oxidation in cells of Acer pseudoplatanus L. grown in culture. Planta 137:71-76
16. KELLY GJ, G ZIMMERMAN, E LATZKO 1976 Light inducted activation of fructose-1,6-bisphosphatase in isolated intact chloroplasts. Biochem Biophys Res Commun 70:193-199

17. KLEPPER L, D FLESHER, RH HAGEMAN 1971 Generation of reduced nicotinamide adenine dinucleotide for nitrate reduction in green leaves. Plant Physiol 48:580-590
18. LANE MD, H MARUYAMA, RL EASTERDAY 1969 Phosphoenolpyruvate carboxylase from peanut cotyledons. Methods Enzymol 24:277-283
19. LARSEN PO, KL CORNWELL, SL GEE, JA BASSHAM 1981 Amino acid synthesis in photosynthesizing spinach cells. Plant Physiol 68:1231-1236
20. LAWYER AL, LK CORNWELL, PO LARSEN, JA BASSHAM 1981 Effects of carbon dioxide and oxygen on the regulation of photosynthetic carbon metabolism by ammonia in spinach mesophyll cells. Plant Physiol 68:292-299
21. LILLEY R McC, DA WALKER 1974 An improved spectrophotometric assay for ribulose bisphosphate carboxylase. Biochem Biophys Acta 358:226-229
22. MAGALHAES AC, CA NEYRA, RH HAGEMAN 1971 Nitrite assimilation and amino nitrogen synthesis in isolated spinach chloroplasts. Plant Physiol 53:411-415
23. MARQUES IA, MJ OBERHOLZER, KH ERISMANN 1983 Effects of different inorganic nitrogen sources on photosynthetic carbon metabolism in primary leaves of non-nodulated Phaseolus vulgaris L. Plant Physiol 71:555-561
24. MONDAL MH, WA BRUN, ML BRENNER 1978 Effects of sink removal on photosynthesis and senescence in leaves of soybean (Glycine max L.) plants. Plant Physiol 61:394-397
25. OLIVER DJ, JH THORNE, R POINCELOT 1979 Rapid isolation of mesophyll cells from soybean. Plant Sci Lett 16:149-155
26. PAUL JS, KL CORNWELL, JA BASSHAM 1978 Effects of ammonia on cell metabolism in photosynthesizing isolated mesophyll cells from Papover somniferum L. Planta 142:49-54
27. PLATT SG, Z PLAUT, JA BASSHAM 1977 Ammonia regulation of carbon metabolism in photosynthesizing leaf discs. Plant Physiol 60:739-742
28. PLAUT Z, K LENDZIAN, JA BASSHAM 1977 Nitrite reduction in a reconstituted and whole chloroplasts during carbon reduction. Plant Physiol 59:184-188

29. PREISS J 1984 Starch, sucrose biosynthesis, and carbon partitioning in plants. Trends in Biochem Sci 9:24-27
30. REED JA, DT CANVIN 1983 Light and dark controls of nitrate reduction in wheat (Tricicum aestivum L.) protoplasts. Plant Physiol 69: 508-513
31. REED JA, DT CANVIN, JH SHERRARD, RH HAGEMAN 1983 Assimilation of 15N nitrate and 15N nitrite in leaves of five plant species under light and dark conditions. Plant Physiol 71:291-294
32. ROBINSON JM 1984 Photosynthetic carbon metabolism in leaves and isolated chloroplasts from spinach plants grown under short and intermediate photosynthetic periods. Plant Physiol 75:397-409
33. ROBINSON JM, FW SNYDER 1983 Influence of N acquisition and assimilation upon partitioning of newly synthesized C-photosynthate in mature soybean leaves. Plant Physiol 72:S-42
34. SARKISSIAN GS, MW FOWLER 1974 Interrelationships between nitrate assimilation and carbohydrate metabolism in plant roots. Planta 105:15-24
35. SCHOLL RL, JE HARPER, RH HAGEMAN 1974 Improvements of nitrite color development in assays of nitrate reductase by phenazine methosulfate and zinc acetate. Plant Physiol 53:825-828
36. THORNE JH, HR KOLLER 1974 Influence of assimilate demand on photosynthesis, diffusive resistances, translocation, and carbohydrate levels of soybean leaves. Plant Physiol 54:201-207
37. WALLSGROVE RM, AJ KEYS, P LEA, BJ MIFLIN 1983 Photorespiration and nitrogen metabolism. Plant, Cell, Environ 6:301-309
38. WOO KC, DT CANVIN 1980 Effect of ammonia on photosynthetic carbon fixation in isolated spinach leaf cells. Can J Bot 58:505-510
39. WOO KC, DT CANVIN 1980 Effect of ammonia, nitrate, and inhibitors of N metabolism on photosynthetic carbon fixation in isolated spinach leaf cells. Can J Bot 58:511-516
40. WOO KC, CB OSMOND 1982 Stimulation of ammonia and 2-oxo-glutarate dependent O_2 evolution in isolated chloroplasts by dicarboxylates and the role of the chloroplast in photorespiratory nitrogen recycling. Plant Physiol 69:562-596

41. YEMM EW, EC COCKING 1955 The determination of amino acids with ninhydrin. Analyst 80:209-213
42. YEMM EW, AJ WILLIS 1956 The respiration of barley plants. IX. The metabolism of roots during the assimilation of nitrogen. New Phytol 55:229-252

Diurnal Changes in Carbon Allocation: Morning

BERNADETTE R. FONDY and DONALD R. GEIGER

Allocation of carbon recently fixed to a number of alternative uses in source leaves is an important means for controlling export, and can be a determinant of crop yield. In sugar beet, there are adjustments in the proportion of newly fixed carbon allocated to sucrose and starch at the beginning and end of the day; yet during most of the day, distribution to these compounds is steady under a range of conditions. These patterns of carbon allocation can provide evidence for understanding adaptive priorities for carbon use during the diurnal cycle.

Recently, investigators have proposed a number of mechanisms for regulation of carbon allocation to sucrose and starch synthesis. These regulatory processes are thought to coordinate sucrose synthesis rate with net carbon export rate, to control storage of sucrose in the exporting leaf, and to moderate sucrose formation, allowing starch synthesis even in the presence of rapid synthesis and use of sucrose. It is important to characterize the mechanisms for controlling these critical physiological priorities under usual circumstances.

In this work, we studied carbon allocation in exporting sugar beet leaves to identify priority uses for recent products of photosynthesis and to understand regulatory mechanisms operating at the beginning of the day.

MATERIALS AND METHODS

A multigerm variety of Beta vulgaris (L.) was raised in a 14-h light period (450 µmol photons m^{-2} s^{-1} PAR) and used when 6- to 8-weeks-old.

Synthesis rates for starch and sucrose were estimated by pulse labeling with $^{14}CO_2$ for 3 min to minimize turnover. Cups were attached to 1.7-cm^2 areas of leaves. A 0.1-ml mixture of 350 µl L^{-1} CO_2 with 2 µCi of $^{14}CO_2$ was injected into each cup which contained 5 ml of air stirred with a stir bar. After each pulse, a 0.78-cm^2 area of leaf was excised and plunged into a hot 1:4 chloroform:methanol (v/v), then extracted in 80% ethanol. Extracts containing the soluble components of the leaf were analyzed for ^{14}C-labeled and total sucrose. The extracted disc was analyzed for ^{14}C-labeled and total starch.

Specific radioactivity of the fixed gas was determined from the net carbon export data, obtained the day before, and the ^{14}C fixed by the leaf area rather than from the gas supplied. This specific radioactivity and the radioactivity in the respective compounds was used to determine the rate of synthesis of sucrose and starch from the recently fixed $^{14}CO_2$.

Mobilization of $^{14}CO_2$ from recently synthesized starch was studied using steady-state labeling. During a 14-h photoperiod, the leaf in a closed system was supplied $^{14}CO_2$ maintained at 380 µl L^{-1} CO_2 with an average specific radioactivity of 1.2 nCi µg^{-1}C. Samples were removed from the leaf 5 min before the end of the light period, and at intervals during the last 2 h of the dark period and the first 5 h of the next light period. The leaf was kept in room air during this second light period.

RESULTS

A comparison of cumulative net carbon export, obtained by integrating values for net carbon export rate, and starch storage showed that although net carbon export rate was at its highest point and recently fixed carbon accumulated in a nearly linear manner early in the light period, starch accumulation remained near zero. During the first 2-h period, a gradual acceleration in starch accumulation closely paralleled the decline in sucrose storage. This suggests that carbon allocated to storage may be diverted to sucrose because of delayed activation of the starch-synthesis pathway following the night.

Evidence of active synthesis pathways was investigated by pulse labeling with $^{14}CO_2$. To obtain initial synthesis rates for sucrose and starch rather than steady-state levels, the rate of entry of ^{14}C into these compounds was measured by 3-min pulses during the beginning phase of the light period. The data revealed changing initial synthesis rates for both sucrose and starch early in the light period, with an increase in sucrose synthesis corresponding to the increase in photosynthesis rate. Synthesis of starch from newly fixed carbon began within minutes after lights were turned on, reaching a value of 60% of the midday rate within 20 min. This rate should have resulted in an increase in starch of 10 μ gC cm^{-2} after 60 min in the light. The increase actually took 120 min, suggesting a degradation of newly made starch in addition to slower synthesis.

Steady-state labeling was used to test the hypothesis that lessening starch accumulation resulted from starch turnover. Loss of ^{14}C from starch was followed at the end of the night, and the beginning of the day in a leaf that had been supplied with $^{14}CO_2$ for 14 h the previous day. ^{14}C was lost from the starch pool during the following dark period, but not after the initiation of the next light period. In contrast, ^{14}C was lost from sucrose throughout the period studied. This loss was expected for a translocating plant.

The influence of increased supply or decreased use of sucrose on starch accumulation was studied to see if carbon demand for sucrose synthesis might have caused starch to turnover. Response of starch accumulation to blocked translocation was compared with that of a translocating leaf. Girdling the petiole of one leaf just prior to the beginning of the light period caused sucrose to increase in the leaf, but did not shorten the lag in starch accumulation. This suggests that carbon needed for sucrose synthesis does not cause starch to turnover.

DISCUSSION

It seems likely that storage of sucrose and starch constituted alternative uses of most of the newly fixed carbon not exported. This interpretation was borne out by the fact that the amount of sucrose accumulated was similar to the deficit in starch storage. Carbon allocated to storage appeared to be regulated to give a steady amount throughout the day even though distribution between the two forms changed.

At the start of the day, there appeared to be an initial priority for storage of a certain amount of sucrose and a resulting diversion of carbon from starch storage. This storage early in the day may have been in response to a need to restore vacuolar sucrose depleted during the night (1). Rates of synthesis determined by pulse-labeling confirmed that sucrose synthesis was more rapid at the time sucrose accumulated during the first hours of the day, and slowed concurrently with the gradual cessation of sucrose accumulation. Results from the study of turnover show that the bulk of the starch was not mobilized to provide carbon for sucrose synthesis, but turnover of the most recently synthesized starch could not be ruled out.

As sucrose reached a critical level, ascendancy of control seemed to shift toward a steady rate of starch accumulation. Increased carbon demand for sucrose synthesis may have caused breakdown of the most recently made starch, resulting in a lag in storage. Starch synthesis with little or no net accumulation of starch lends support to simultaneous synthesis and degradation of recently made starch. It is not clear why slowing the use of sucrose for export did not result in earlier accumulation of starch.

LITERATURE CITED

1. KAISER G, E MARTINOIA, A WIEMKEN 1982 Rapid appearance of photosynthetic products in the vacuoles isolated from barley mesophyll protoplasts by a new fast method. Z Pflanzenphysiol 17:103-113

Nonvascular Transfer of Assimilates in Citrus Juice Tissues

KAREN E. KOCH

Phloem unloading and subsequent transfer of ^{14}C-photosynthates into sink tissues play important roles in the overall translocation process (1). Opportunities to examine these component steps experimentally are rare (1, 4, 5), but the unusual structure of citrus fruit facilitates such investigations. Distinct vascular and nonvascular regions exist, with phloem distribution limited to areas outside juice tissues. The multi-celled juice vesicles are attached to inner walls of the segment epidermis via hair-like stalks, in rows immediately adjacent to the nearest vascular bundles. Only three of these vascular strands are present for a given segment, so phloem-unloading must occur in highly localized areas. Subsequent transfer through phloem-free zones can also be followed in citrus fruit because individual juice vesicle stalks are large enough for separate analysis.

Photosynthates labeled in an individual source leaf are translocated to juice segments in direct vertical alignment (3), and are found primarily in the dorsal vascular bundle along the outer tangential wall of each (2). Kinetic studies have also shown this to be the primary path of photosynthate entry into juice vesicles (unpublished results). Within nonvascular tissues, photosynthates accumulate rapidly in the segment epidermis, but appear only to a limited extent in juice tissues (2). Approximately 70% of the photosynthates going to the latter are localized in vesicle stalks after 6 h of translocation (2).

The long-term transfer of ^{14}C-photosynthates through the nonvascular portion of the transport path in intact fruit is compared in the present report to that of isolated vesicle rows dissected after the first 6 h of translocation.

MATERIALS AND METHODS

Assimilates were photosynthetically labeled in 5- to 7-year-old 'Marsh' grapefruit trees (Citrus paradisi Macf.) by exposing a source leaf nearest or next nearest a fruit to $^{14}CO_2$ for 1 h (2). Translocation in one group of experiments was allowed to continue in ambient air for various periods of time (6, 24, 30, or 48 h). In the other, vesicle rows were isolated from fruit after a 6-h translocation period. Segment epidermis was peeled back from segments in direct vertical alignment to the source leaf and adjacent vesicles teased away. A dorsal vascular bundle was left in association with each row of juice vesicles, and together these were transferred to moist filter paper in sterile petri dishes. Both groups of experiments were terminated by completely separating tissues, freezing them in liquid N_2, and extracting in boiling 80% ethanol (2). Radioactivity was quantified via liquid scintillation procedures.

RESULTS AND DISCUSSION

The change in localization of a 1-h pulse of ^{14}C-photosynthates moving into grapefruit juice vesicles over 48- and 72-h periods is shown in Figure 1. The slow rate of transfer in intact fruit (Fig. 1A) is only partially accounted for by spreading of the initial pulse, because no labeled assimilates moved from the source leaf to the fruit after 30 h, and juice tissues accumulated no further radioactivity from vascular bundles after this point (data not shown). The pulse of labeled assimilates is narrowed in isolated vesicle rows (Fig. 1B) where the influx of ^{14}C-photosynthates from source leaves was terminated by dissection at 6 h. Still, the most radioactive portion of the translocated pulse reaches dorsal vascular bundles prior to this time. Movement of ^{14}C-photosynthates out of the phloem of intact fruit occurs rapidly after the first 6 h, and is complete within 24 h at the stage of fruit development studied here (data not shown).

Redistribution of label from hair-like vesicle stalks to juice vesicles occurred at a linear rate during the time period examined for

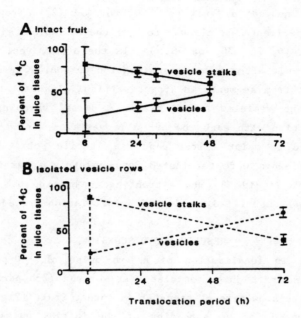

FIG. 1. Change in localization of a 1-h pulse of ^{14}C-photosynthates moving into grapefruit juice vesicles over a 48- to 72-h period. Assimilate redistribution continued in intact, attached fruit (A) until samples were taken at the time intervals indicated in the figure. Vesicle rows with associated vascular bundles (B) were isolated from fruit after an initial transport period of 6 h, and transferred to sterile conditions. Data points for vesicles (●) or vesicle stalks (■) represent means of 4 experiments, except at 6 h in isolated tissues, when radioactivity of individual, isolated vesicle rows was detectable only in vascular strands of 3 experiments. Verticle bars for other points denote standard errors of means.

intact fruit (Fig. 1A). A total of ca. 4.8 d would be required for complete transfer of this pulse to vesicles if the observed rate remained constant. Grapefruit juice tissues do gain measurable dry weight at this stage of development, and 80% of the photosynthates from the nearest source leaf are translocated to whole fruit within 24 h. Subsequent transfer of labeled assimilates through nonvascular tissues after phloem unloading apparently proceeds far more slowly than is observed within vascular structures.

This transfer of labeled assimilates was found to continue in isolated vesicle rows attached to a dorsal vascular bundle (Fig. 1B). Present data are available only for 72-h experiments, so redistribution is not known to proceed linearly over time in the isolated system. The pulse of ^{14}C-photosynthates is also narrower than in intact fruit. Despite these differences, it is apparent that the process does continue in an isolated system. It is therefore also less sensitive to manipulation than is long-distance translocation in phloem.

SUMMARY

Transfer of labeled assimilates through phloem-free areas and into sink tissues was examined in intact grapefruit (Citrus paradisi Macf.) and in rows of juice vesicles isolated with an attached vascular bundle. Entry of ^{14}C-photosynthates into phloem-free juice tissues was completed with 30 h in intact and isolated structures. Subsequent redistribution from hair-like vesicle stalks into juice vesicles, however, continued throughout the experiments. Localization of radioactivity suggests possible similarities in nonvascular transfer rates of both systems.

ACKNOWLEDGMENTS--The author is grateful to W. T. Avigne for skilled technical assistance.

LITERATURE CITED

1. FONDY BR, DR GEIGER 1980 Phloem loading and unloading: Pathways and mechanisms. What's New in Plant Physiology 11:25-28
2. KOCH KE 1984 The path of photosynthate translocation into citrus fruit. Plant Cell and Environ 7:647-653

3. KOCH KE, WT AVIGNE 1984 Localized photosynthate deposition in citrus fruit segments relative to source-leaf position. Plant and Cell Physiol 25:859-860
4. THORNE JH, RM RAINBIRD 1983 An *in vitro* technique for the study of phloem unloading in seedcoats of developing soybean seeds. Plant Physiol 72:268-271
5. WOLSWINKEL P, A AMMERLAAN 1983 Sucrose and hexose release by excised stem segments of *Vicia faba* L.: The sucrose-specific stimulating influence of *Cuscuta* on sugar release and the activity of acid invertase. J Exp Bot 34:1516-1527

Regulation of Carbohydrate Partitioning

HARRY BEEVERS

Carbohydrate partitioning, as reflected by the contributions to this symposium, involves three components and their interaction: the supply of mobile sugars (principally sucrose) by the source leaves, transport to the sink(s), and processing or storage within the sink.

The first requires a full understanding of the carbon metabolism in photosynthetic cells and particularly of those reactions that control and maintain the steady-state level of sucrose in the cytosol in both light and darkness. The history of work in this area has shown a shifting emphasis on what is supposed to be the key regulatory determinant. The rate of sucrose production in leaves is rarely due to the amount of a particular enzyme, except during brief induction phases. For each individual species, of course, there is some upper limit to the rate, but this is rarely achieved under natural conditions. More recent work at the cellular level has emphasized instead the regulation of existing enzyme activities and, at different times, aspects of compartmentation, ion concentrations, pH, and energy charge have been thought to be of overriding importance. The role of specific transporters between chloroplast and cytosol and the possibility that photosynthesis might be limited by their capacity has been raised. Particular emphasis is placed at present on the regulation of key enzyme reactions (those known not to be at thermodynamic equilibrium) by metabolite effectors. In particular, the role of fructose-2,6-bisphosphate and enzymes that in turn regulate its level have received appropriate attention most recently.

At the cellular level then, we are faced with an already wide range of possible regulatory influences on the maintenance of sucrose levels in the cytosol, each of which appears to the particular investigator as the all-important bottleneck. The logic of this approach requires that all enzymes and possible effectors be known, and what the actual levels of those effectors, co-enzymes, and ionic cofactors are in the different cellular compartments. Indeed, some impressive progress has been made. However, when this objective has been achieved, a comprehensive computer analysis of the overall network of pathways and transporters will be required to decide which are, in fact, the important control points. Even then we cannot be confident that we have a final solution. We should acknowledge that the discovery of hitherto unrecognized effectors, feedback metabolites, or hormones from sinks, would modify the conclusions overnight. And the skeptic could always maintain that the determination of (average) concentrations of metabolites in chloroplast or cytosol might not reveal what the actual concentrations are at specific enzyme surfaces within these compartments where localized differences and metabolic channeling might indeed occur.

The transport from photosynthetic cell to sink is itself composed of a series of components. It now appears that sucrose is transported through cytoplasmic connections between photosynthetic cells (the symplast) and moves into the apoplast or free space by a process of facilitated diffusion. It is then apparently loaded into phloem cells by a metabolically-driven process which results in very high concentrations (>500 mM) in the sieve tubes. Although the mechanism of movement in the phloem is not fully understood, it is clear that high rates of transport (100 cm/h) can occur and that unloading is confined to particular points at the phloem extremities with the sugar actively absorbed by the sinks. Each of these steps is a potentially-limiting constraint on the supply of sucrose to sinks, and they are all currently under active investigation.

The classical view that it is the sucrose gradient between source leaf and sink that determines the direction and extent of movement still prevails. Any developing organ or storage tissue represents a sink, and it seems obvious that continued utilization of the entering sugar by any means would maintain the gradient and prolong the import. Competition between potential sinks with established phloem connections has been recognized and the notion of "sink strength" describes this feature, while unfortunately not defining it in quantitative chemical terms. There is some evidence suggesting feedback from sinks to leaves and possible hormonal effects, and it is tempting to conclude that there must be some chemical signals operating between the two compartments. But again, these remain elusive and it seems that it is the establishment and maintenance of sinks that is of major importance.

The presently-increasing interest in the subject of partitioning is based to some extent on the notion that it can be experimentally, possibly for commercial gain, manipulated to increase the yield of that part of the plant which is the desired product. However, although various surgical and environmental treatments have been shown to modify partitioning, there are at present no powerful chemical treatments or cultural practices that ensure the superiority of one sink over another.

It must be acknowledged that the impressive success of generations of plant breeders in producing the high-yielding varieties that we have today has been achieved through persistent selection of morphological types with desirable sink characteristics, and not through deliberate screening for physiological or biochemical features. It has been pointed out, for example, that the present crop species do not have intrinsically higher rates of net photosynthesis than their native progenitors, and it is not yet evident that in the breeding programs, individuals have been serendipitously selected with superior transport characteristics.

Nevertheless, much has been learned in a descriptive sense, in defining possible bottlenecks and control points in the system. Whether, once partitioning is understood more fully, it can be modified for commercial gain, remains an open question. I see no compelling reason for optimism.

INDEX

ADP-glucose, 1, 86
ADP-glucose pyrophosphorylase, 1, 7, 37, 120, 151, 347
Allium cepa, 81, 259
Allium porrum, 81
α-glucose-1-phosphate, 7
Amino acid metabolism, 347-351
Ammonium ion (NH_4^+), 337
AMP, 63
Amylase, 16, 28, 35-36
Amylolysis, 27-42
Amylopectin, 1, 4, 28
Amyloplast, 83, 142
Amylose, 4, 28
Anaplerotic pathway, 335, 345
Apium graveolens, 312
Apoplastic space, 232
Arum maculatum L., 78-88
Assimilatory starch, 28, 30
ATP, 7, 157, 236
ATPase, 106, 232

β-(2-1)fructan:β-(2-1)fructan 1-fructosyl transferase, 277

β-limit dextrin, 28
Beta vulgaris, 6, 258, 289-304, 358
Branching enzyme, 4, 37
Bundle sheath cells, 17, 114, 128

C_3 plants, 15
C_4 plants, 17, 128
Calcium, sucrose transport, 219
CCCP, 236
Chlorophyll fluorescence, 93-106
Chloroplasts, 184-188, 192-197
Chloroplast envelope transport, 147-159
Chloroplast DNA, 128-131
Chloroplasts, reconstituted, 197, 335
Chloroplast, starch synthesis, 152
Chromatofocusing, 65
Citrus paradisi Macf., 363
Clarkia, 132
CO_2 compensation point, 313-328
Commelina communis, 13

Compartmentation, 63, 109-114,
Cytochrome C oxidase, 15

Day length and starch, 289-30
Debranching enzyme, 16, 36
Diffusion gradients, PGA, DHAP, 114
Digitaria pentzii, 18
Dihydroxyacetone phosphate (DHAP), 111-115, 147-158, 200
Diurnal measurements, 56, 201-212, 289-304, 358-361
Dulcitol, 309

Enolase, 142
Enzyme "controllability", 122
Enzyme interconversion, 74, 78, 134
Escherichia coli, 216, 235

Fatty acids,
Ferredoxin, 342
Ferredoxin-NADP reductase, 343
Floral development, 283
Fluorescence oscillations, 93-106
Fraxinus, 312
Fructan, 274-286
Fructose, 87-88, 262
Fructose-1,6-bisphosphatase (FbPase) 58, 63-74, 111, 148, 199, 319, 345
Fructose-1,6-bisphosphate, 65, 200
Fructose-1,6-bisphosphate aldolase, 84
Fructose-2,6-bisphosphate, 45-59, 77, 109-124, 200, 206-208, 367

Fructose-6-P, 48, 80, 118 127-143, 367-369
Fructosyl transferase, 277
Gas transients, 93-106
Glucan:glucose glucosyl transferase, 27, 39
Glucose-1-P, 36, 88
Glucose-6-P, 88, 118
Glucose-6-P-dehydrogenase, 345
Glutamate synthase (barley mutant), 100
Glyceraldehyde-3-P dehydrogenase, 84, 137, 189, 311
Glycogen, 34
Grapefruit juice vesicles, 362-365
Growth rates (diurnal), 208-209, 295-304
Guard cells, 13, 16, 162-179

Hexokinase, 40, 101
Hexose phosphate isomerase, 132
Hordeum vulgare, 6
Hoya carnosa, 6

Inulin, 276-278
Invertase, 85-87, 131, 278
Ionophores, 227, 235-236
Isoelectric focusing, 63
Isozymes, 127-143

Lactuca sativa, 6
Leucoplast, 127-143
Leucoplast DNA, 128-131
Light-scattering, 98-100
Lolium temulentum, 81

Lycopersicon esculentum, 6

Maltase, 35
Maltodextrins, 37
Maltose, 37
Maltose phosphorylase, 27
Malus, 310
Manihot esculenta, 6
Mannitol, 236, 309-323
Mannitol-1-P -phosphatase, 311
Mannose "feeding", 98-105
Mannose phosphate, 102
Mannose-6-P reductase, isomerase 311, 325
Membrane potential, 232, 267
Mesophyll, 15, 17, 114, 128, 257, 335
3-O-Methyl glucose, 257
Mg^{++}, 7, 65
"MICROENZYME" computer program, 168-176
Models: kinetic, partitioning, 148-151, 291-293
Monoclonal antibodies, 3

Net carbon exchange (NCE), 200, 290-302
Nicotiana tabacum, 6
Nitrate (-ite), 337-346
Nitrite reductase, 337-346
Nitrogen assimilation, 333-353
Non-structural carbohydrates (NSC) 274-286
Nuclear coding, isozymes, 131

Oligoglucans, 37
Oryza sativa, 6
Oxidative phosphorylation, 15
Oxygen evolution, 94

PAGE, 68, 132
PCMBS, 244
Pentose phosphate pathway, 40, 63, 88, 311, 345
PEP carboxylase, 15, 165-168, 319, 345
PGA/Pi ratios, 6, 120, 147-159, 347
pH, sucrose uptake, 217-220, 367
Phaseolus vulgaris, 6
Phleins, 276-278
Phloem (un)loading, 232, 254-269, 362-265, 368

Phosphate, 6, 20, 98, 147-159
Phosphofructokinase (PFKase), 46, 76-106, 80, 132-137
Phosphofructophosphotransferase (PFPase), 46-59, 76-106, 111, 132-137
Phosphogluconate dehydrogenase, 127
6-Phosphogluconate dehydrogenase, 345
Phosphoglyceric acid (PGA), 6, 114, 132, 147-158, 347
Phosphoglyceromutase, 138-139
Phosphorolysis, 27-42
Phosphorylase, 16, 18, 27, 36
Photoinhibition, 181, 311

Photoperiod, 122, 294
Photophosphorylation, 13
Photorespiration, 180, 190, 312, 315
Photosynthesis measurements, 93-106, 152-157, 181, 192-197, 203, 313, 333-353, 360
Photosystem II, 95, 114
Pi/DHAP ratios, 148-158
Pi/PGA ratio, 148-158
Pineapple PFPase, 46-51
Ping-pong mechanism, 46
Pisum sativum, 13, 77, 86-87
Plantago coronopsis, 327
Plastoquinone, 95
Polyglucans, 37
Polyols (sugar alcohols), 309-328
Potato tubers, 86-87
Proplastid, 128
Proton gradient, 97, 98, 105
Proton-motive force, 222, 232
Proton "well", 225
Prunus, 310
Pullulan, 38
Pyridine nucleotides, 311, 333
Pyrophosphatase, 7, 54, 79
Pyrophosphate, 7, 12, 45-59, 76-88, 98, 147-159
Pyrus, 310, 312
Pyruvate kinase, 345

"Q", 95, 97, 105
Quartz fiber balance, 165-166

R-enzyme, 28, 35

Respiration, 80-85
Ribose-5-P isomerase,epimerase, 188-190
Ribulose-1,5-bisphosphate (RuBP) 110, 151, 181, 185, 190
Ribulose-1,5-bisphosphate carboxylase (Rubisco) 14, 138, 180, 185, 190, 319, 345
Ricinus, 255, 265
Root:Shoot ratio, 289-304
Roots, respiration in, 81

Sieve element-companion cell (se-cc) complex, 255, 264
Single cell enzyme assays, 162-179
Sorbitol, 309-312
Sorbitol dehydrogenase, 311
Sorbitol oxidase, 311
Sorbitol-1-P, 311
Sorghum dochma, 6
Source-sink relations, 192-212, 232, 273, 289-304, 312, 339, 367-369
Spinacea oleracea, 6, 184
Starch accumulation, 109, 247, 358-361
Starch degradation, 27-42
Starch synthase, 1, 2, 37
Starch synthesis, 1-20, 109-124, 152, 359-361
Streptococcus lactis, 215
Streptococcus mutans, 278
Sucrose, 40, 85-88, 231, 273-286, 300, 314, 367-369

Sucrose glycolysis, 76-88
Sucrose phosphate synthase, 114, 118, 122, 148, 199-212, 314
Sucrose synthase, 85-88, 314
Sucrose synthesis, 109-124, 199-212, 359-361
Sucrose uptake/leakage, 215-229, 236-251, 277-282
Sucrose-H+ cotransport, 215-229, 256-269
Synechococcus leopoliensis, 63-71

Temperature coefficients, photosynthesis, 193
Thermogenesis, 82
Tonoplast transport, 235
Transglycosidase, 36
Transitory starch, 40
Transketolase, 188-190, 345
Translocation, 199-212, 254-269, 309-328, 362-365
Triose P/phosphate translocator, 15, 20, 40, 110, 132, 147-151
Triose phosphate, 15, 102, 110, 132
Triticum aestivum, 6
Turgor, 208, 241, 273

UDP-glucose, 1, 40, 86, 118
UDP-glucose pyrophosphorylase 87-88, 134

Vicia faba, 14, 131, 163-165
Vitis vinifera, 6

Winter wheat, 273-286

Xerosicyos danguyi, 6
Xylulose-5-P, 188

Zea mays, 6, 216, 264